Circular Economy and Fly Ash Management

Sadhan Kumar Ghosh
Vimal Kumar
Editors

Circular Economy and Fly Ash Management

Springer

Editors
Sadhan Kumar Ghosh
Department of Mechanical Engineering
Jadavpur University
Kolkata, West Bengal, India

Vimal Kumar
Centre for Fly Ash Research
and Management
New Delhi, Delhi, India

ISBN 978-981-15-0013-8 ISBN 978-981-15-0014-5 (eBook)
https://doi.org/10.1007/978-981-15-0014-5

This Springer imprint is published by the registered company Springer Nature Singapore Pte Ltd.
The registered company address is: 152 Beach Road, #21-01/04 Gateway East, Singapore 189721, Singapore

Preface

The generation, management, treatment and disposal of fly ash all over the world have been regarded as serious issues of solid waste. Fly ash is a coal combustion residue of coal-/lignite-based thermal power plants (C/LBTPS). Global energy demand is set to increase by almost 50% in the period 2016–2040. Coal, the most abundant fossil fuel on the planet, is relatively cheap with some of the largest deposits in regions that are relatively stable politically, such as China, India and the USA. In the last half-century, coal has been a dominant player in energy generation worldwide and is projected to maintain its dominance in decades to come. Much of this growth will continue to be concentrated in the developing world, primarily China and India, and will propel the need for energy in general and coal in particular. India is the third largest producer of coal. Indian coal has high ash content, nearly 30–45%, and produces a large quantity of fly ash at C/LBTPS. Nearly 73% of India's total installed power generation capacity is thermal, of which coal-based generation is 90%. Nearly 35–40% fly ash in bituminous or sub-bituminous coal remains unutilized in India. Various estimates indicate that electricity generated from coal is expected to grow twofold to threefold by 2030. To meet the growing energy demand and thereby increase the power-generating capacity, the dependency on coal for power generation and disposal of fly ash will continue to increase along with various unavoidable problems.

Due to the physical characteristics and sheer volumes generated, serious problems with fly ash in several aspects include: (1) fly ash particles both as dry ash and as pond ash occupy many hectares of land in the vicinity of power station due to heavy disposal; (2) because of its fineness, it is very difficult to handle fly ash in dry state, and flying fine particles of ash corrode structural surfaces and affect horticulture; (3) it disturbs the ecology through soil, air and water pollution; (4) long inhalation of fly ash causes various serious diseases like silicosis, fibrosis of lungs, bronchitis and pneumonitis; and (5) the oxides of iron and aluminium present on the surface of the fly ash particles attract toxic trace elements, e.g. Sb, As, Be, Cd, Pb, Hg, Se and V, and they are found to be concentrated largely on the surface of fly ash.

The Ministry of Environment, Forest and Climate Change (MoEFCC), Government of India, has prescribed the targets for fly ash utilization for C/LBTPS to achieve 100% utilization in a phased manner driven by the circular economy and 5 Rs concepts. Central Electricity Authority has been monitoring the status of fly ash generation and its utilization since 1996. A large number of technologies have been developed for gainful utilization and safe management of fly ash under the concerted efforts made by Fly Ash Mission/Fly Ash Unit under the Ministry of Science and Technology, Government of India, since 1994. As a result, fly ash earlier considered to be "hazardous industrial waste" material has now acquired the status of useful and saleable commodity. The utilization of fly ash has increased from 6.64 million ton in 1996–1997 to a level of 107.10 million ton in 2016–2017. The percentage of fly ash utilization during 2016–2017 is 63.28%. It is, however, noted that 100% utilization of fly ash on all India basis is unlikely to materialize. As per the report of the Central Electricity Authority (CEA), Government of India, during the year 2016–2017, 50 thermal power stations have achieved the fly ash utilization level of 100% or more including 39 thermal power stations which have achieved fly ash utilization level of more than 100%. Fly ash utilization level of more than 100% indicates that besides ash generation during the period of report 2016–2017, additional fly ash from ash pond was utilized. Nearly 730 tons of coal is consumed by the power plants tuning to an average of 210 tons of fly ash to generate in 2018, whereas nearly 140 tons of fly ash per annum is being utilized in India resulting in reduction in 75 tons of CO_2 generation by cement and brick industries with a business worth Rs. 100 billion a year, creating 1.5 million of employment.

The potential applications for coal fly ash as a raw material include soil amelioration agent in agriculture, highway embankments, construction of bricks, aggregate material in Portland cement, filling of low-lying areas, manufacture of glass and ceramics, production of zeolites, formation of mesoporous materials, synthesis of geopolymers, catalysts and catalyst supports, adsorbent for gases and wastewater processes, and extraction of metals. CEA's report indicates that the utilization of fly ash is the highest in the cement sector at 24.04%, followed by bricks and tile industries at 7.37% and the concrete industry segment at the lowest level of utilization at 0.6% in India. Among the section of utilization in China as per the National Development and Reform Commission (NDRC), the top three were cement (41%), brick and tiles (26%) and concrete (19%). The coal ash by-product has been classified as a green list waste under the Organisation for Economic Co-operation and Development (OECD). Coal ash by-product is not considered as a waste under Basel Convention, whereas in many countries this industrial by-product has not been properly utilized, rather it has been neglected like a waste substance. Thus, fly ash management is a cause of concern for the present and future. There is a trend worldwide to circulate the potential resources to achieve the Sustainable Development Goals. Many countries are trying to implement the concept of circular economy to achieve the resource-efficient system supported by country policies. The management and utilization of fly ash are the part of the implementation of circular economy and 5 Rs (reduce, reuse, recycle, remanufacture and repair) concepts in India, China and many other countries.

The 8th IconSWM 2018 received 380 abstracts and 320 full papers from 30 countries. Three hundred accepted full papers have been presented in November 2018 at Acharya Nagarjuna University, Guntur, Andhra Pradesh, India. After a thorough review by experts and required revisions, the board has finally selected eleven chapters for this book, *Circular Economy and Fly Ash Management*, dealing with the utilization of fly ash as a replacement for chemical pesticides, in vermicomposting and in agriculture, carbon and nutrient sequestration potential, fly ash-based herbal pesticides in agriculture, household, poultry and grains in storage, assessment of fly ash-based chemical pesticides, fly ash nanoparticle technology in pest control, behaviour and strength of fly ash concrete, fly ash as green technology inputs and fly ash as a source of silicon for mitigating biotic stress.

The IconSWM movement was initiated focusing on better waste management, resource circulation and environmental protection since the year 2009. It helps generating awareness and bringing all the stakeholders together from all over the world under the aegis of the International Society of Waste Management, Air and Water (ISWMAW). It established a few research projects across the world involving CST at the Indian Institute of Science, Jadavpur University and a few other institutions in India and experts from more than 30 countries in the research project on circular economy. Consortium of Researchers in International Collaboration (CRIC) and other organizations across the world are helping the IconSWM movement. IconSWM has become one of the biggest platforms in India for knowledge sharing and awareness generation among the urban local bodies (ULBs), government departments, researchers, industries, NGOs, communities and other stakeholders in the area of waste management. The primary agenda of this conference is to reduce the waste generation encouraging the implementation of 5 Rs (reduce, reuse, recycle, remanufacture and repair) concept. The conference provided holistic pathways to waste management and resource circulation conforming to urban mining and circular economy.

The success of the 8th IconSWM 2018 is the result of effective contribution of the government of Andhra Pradesh, several industry associations, chamber of commerce and industries, AP State Council of Higher Education, various organizations and individuals in India and abroad. Support of UNEP, UNIDO, UNCRD, delegation from European Union and other foreign organizations were significant. The 8th IconSWM 2018 was attended by nearly 823 delegates from 22 countries. The 9th IconSWM 2019 will be held at KIIT, Bhubaneswar, Odisha, during 27–30 November 2019, and we are expecting nearly 900 participants from 30 countries.

This book will be helpful to the educational and research institutes, policymakers, government, implementers, ULBs and NGOs. Hope to see you all in the 9th IconSWM-CE 2019 in November 2019.

Kolkata, India
November 2019

Prof. Sadhan Kumar Ghosh
Dr. Vimal Kumar

Acknowledgement

The Hon'ble Chief Minister and Hon'ble Minister of MA&UD for taking personal interest in this conference.

We are indebted to Shri. R. Valavan Karikal, IAS; Dr. C. L. Venkata Rao; Shri. B. S. S. Prasad, IFS (retd.); Prof. S. Vijaya Raju; and Prof. A. Rajendra Prasad, VC, ANU, for their unconditional support and guidance for preparing the platform for the successful 8th IconSWM 2018 at Guntur, Vijayawada, AP.

I must express my gratitude to Mr. Vinod Kumar Jindal, ICoAS; Shri. D. Muralidhar Reddy, IAS; Shri. K. Kanna Babu, IAS; Mr. Vivek Jadav, IAS; Mr. Anjum Parwez, IAS; Mr. Bala Kishore; Prof. S. Varadarajan; Mr. K. Vinayakam; Prof. Shinichi Sakai, Kyoto University, JSMCWM; Prof. Y. C. Seo and Prof. S. W. Rhee of KSWM; Shri. C. R. C. Mohanty, UNCRD; members of Industry Associations in Andhra Pradesh; Prof. P. Agamuthu, WM&R; Prof. M. Nelles, Rostock University; Dr. Rene Van Berkel, UNIDO; and Ms. Kakuko Nagatani-Yoshida and Mr. Atul Bagai of UNEP and UN delegation to India for their active support.

IconSWM-ISWMAW Committee acknowledges the contribution and interest of all the sponsors, industry partners, industries, co-organizers, organizing partners around the world; the government of Andhra Pradesh; Swachh Andhra Corporation as the principal collaborator; the vice chancellor and all the professors and academic community at Acharya Nagarjuna University (ANU); the chairman, vice chairman, secretary and other officers of AP State Council of Higher Education for involving all the universities in the state; the chairman, member secretary and the officers of the AP Pollution Control Board; the director of factories, the director of boilers, the director of mines and the officers of different ports in Andhra Pradesh; and the delegates and service providers, for making 8th IconSWM 2018 a successful event.

I must specially mention the support and guidance by each of the members of the International Scientific Committee, CRIC members, the core group members and the Local Organizing Committee members of 8th IconSWM 2018 who are the pillars for the success of the programme. The editorial board members including the reviewers, authors and speakers and Mr. Aninda Bose and Ms. Kamiya Khatter of

M/s. Springer India Pvt. Ltd deserve thanks who were very enthusiastic in giving me inputs to bring this book.

I must mention the active participation of all the team members in IconSWM Secretariat across the country with special mention of Prof. H. N. Chanakya and his team in IISc Bangalore; Ms. Sheetal Singh and Dr. Sandhya Jaykumar and their team in CMAK & BBMP; Mr. Saikesh Paruchuri, Mr. Anjaneyulu, Ms. Senophiah Mary, Mr. Rahul Baidya, Ms. Ipsita Saha, Mr. Suresh Mondal, Mr. Bisweswar Ghosh, Mr. Gobinda Debnath and the research team members in Mechanical Engineering Department and ISWMAW, Kolkata HQ, for various activities for the success of the 8th IconSWM 2018. I express my special thanks to Sannidhya Kumar Ghosh, being the governing body member of ISWMAW supported the activities from the USA.

I am indebted to Mrs. Pranati Ghosh who gave me guidance and moral support in achieving the success of the event. Once again, IconSWM and ISWMAW express gratitude to all the stakeholders and delegates, and speakers who are the part of the success of the 8th IconSWM 2018.

Contents

About the Editors

Prof. Sadhan Kumar Ghosh, PhD Professor in mechanical engineering and founder coordinator of Centre for quality management system at Jadavpur University, India. He was the Director of CBWE, Govt. of India and served in Larsen and Toubro Ltd. He is internationally renowned personality in the field of Waste Management, Green Manufacturing, Supply Chain Management, Circular economy, and ISO standards. He has more than 180 research publications in leading international refereed journals, conference proceedings and books having edited more than 30 books and proceedings. His research collaboration exists directly in more than 30 countries. He planted more than 15,000 trees in India whose age ranges from 1 year to 36 years, surviving nearly 40 %.

He has accomplished and has been involved in several impactful interdisciplinary research projects on sustainable supply chain of small and medium sized enterprises (SMEs) across the globe, Circular Economy in 34 countries (2018-2021), waste to energy, e-waste management in BRICS and other related areas. His projects have been funded by European Union Horizon 2020 (2018-2022), Royal Academy of Engineering (2018-2020 & 2012), Shota Rustaveli National Science Foundation (SRNSF) of Georgia (2019-2021), UNCRD/DESA as Expert (2016-2018), Asian Productivity Organisation (APO) (2016-2019), British Council & DST (2012-2014), Royal Society (2015), Erasmus Plus (2016-17), ISWMAW (2018-21), Jute Technology

Mission (2008-2011), Central Pollution Control Board (1999-2002), Govt. of India, and a few others.

He is currently facilitating Circular Economy implementation in several countries and in SMEs in India and the UK to adopt low carbon initiatives and achieve competitiveness. He is the Associate editor of Journal of Japan Society of Materials Cycles and Waste Management. He has been invited for key note and plenary speeches in many countries across the globe. He was the International Expert/Consultant of UN DESA/UNCRD in 2016-18 and APO, Japan during 2014-19. He works as expert in many government committees in India, Japan, Asia Pacific and in South Asian countries. He received Indian patent on "Eco-friendly plastics recycling machine and the process there of" [no.202532 dt 02/03/2007] and on "Automatic High Speed Jute Ribboning Machine", Bangladesh patent no. 1005146 dt 17/02/2014] and Indian Patent no. 306869 dt 05/02/2019. He is associated with ISO Working Groups and BIS concerning waste management (ISO/TC 297) & CHD 34 concerning environment management (ISO/TC 61 and ISO TC 207). He is a lead assessor in ISO 9001, ISO 14001, OHSAS 18001, HACCP 22001 and trainer and auditor in these areas of International standards on management systems. He wrote nine books and a number of book chapters. He is available at : sadhankghosh@gmail.com and website : www.sadhankghosh.com.

Dr. Vimal Kumar PhD Honorary Secretary General, C-FARM is a Mechanical Engineer with MBA. He is the founder Mission Director, Fly Ash, Department of Science & Technology (DST) for implementation of the Fly Ash Mission and turned around the image of fly ash from "a polluting industrial waste" to "a resource material". He developed National Guideline for Re-use and Recycling of C&D Waste including indicative feasibility report for 8 cities and DPRs for 2 cities for Building Materials and Technology Promotion Council (BMTPC), Ministry of Housing and Urban Affairs (MoHUA), Government of India. He has also been associated with Municipal Corporations of Surat, Greater Hyderabad and Greater Mumbai for setting up of C&D waste processing and re-use facilities and spearheaded the development of mobile plant for in-situ processing of C&D waste with BMTPC, MoHUA, Government of India and industries. Dr. Kumar led the Joint Working Group of Indian in shared expertise and technologies of Fly Ash with Russian Federation. His research interest includes waste management, environment protection, development and commercialization of new technologies, technology forecasting & assessment and new construction technologies. He has traveled widely across the globe, published/presented more than 150 papers, been a visiting faculty to renowned management and technology institutes, member of Governing Councils & Research Bodies of a number of R&D institutes and on the Board of Editors of International Journal of Technology Transfer and Commercialization (IJTTC), UK. He has co-ownership to six patents. He is the Chairman and member of several expert committees at state and central government. He is the Governing Council member of Indian Building Council, New Delhi, Member in BIS Committees and Consultant, World Bank.

Handling and Utilisation of Fly Ash from Thermal Power Plants

S. A. Nihalani, Y. D. Mishra and A. R. Meeruty

Abstract Electric power in India mainly depends on coal-fired power plants. Commonly, Indian coal comprises ash in the range of 30–45%. In order to sustain India's economic growth, the country total coal demand is forecasted to more than double by 2030. Increasingly huge quantities of fly and bottom ash are produced in the country, thus leading to the necessity to duly plan safe and clean ways to handle, use, and dispose of the combustion by-products. Thermal power plant design nowadays must duly consider apprehensions related to water shortage, environmental guidelines, sustainable management and disposal of ash, along with growing consciousness pertaining to overall cost and power plant efficiency. The current paper discusses the problems associated with fly ash and its handling and mitigation measures. Fly ash generated while burning of coal in thermal power plants can be utilised for several favourable uses like manufacturing of cement, road construction, road embankment and development of ceramics or fertiliser.

Keywords Suspended particulate emission control · Electrostatic precipitator · Fly ash handling

1 Introduction

India is amongst the largest economies of the world and has a quick growing energy market. The principal sources of energy are coal and petrol in India. The per capita availability and utilisation of electricity are very less in India as compared to the developed world. The rapid economic growth and the resultant increased standard of living of the population call for the huge increase in the supply of power. Latest technological innovations need to be adopted by the existing as well as forthcoming power plants not only to increase the efficiency of power production but also to increase the life of the power plants. However, a higher percentage of ash and incompetent

S. A. Nihalani (✉) · A. R. Meeruty
Department of Civil Engineering, Parul University, Vadodara 391 760, India
e-mail: seemanihalani@yahoo.com

Y. D. Mishra
L&T-Sargent & Lundy Limited, Vadodara, India

S. K. Ghosh and V. Kumar (eds.), *Circular Economy and Fly Ash Management*,
https://doi.org/10.1007/978-981-15-0014-5_1

incineration technologies lead to the discharge of particulates and various harmful gases in the atmosphere. These pollutants are responsible for the greenhouse effect of climate change and various other air pollution problems. Fly ash being essential component present in the coal further worsens the problem of handling and disposing it.

When the combustion of coal occurs, the fine particles flushed from the boiler consist of fly ash along with the flue gases. In this process, the portion of ash that is accumulated at the boiler bottom is termed as bottom fly ash. Handling of fly ash and its recovery and reuse is becoming a growing apprehension day by day, on account of rising landfill costs and recent importance in sustainability. Only about 50–60% of the total fly ash produced by the power plants is currently being reused while the residual is discarded in ash ponds. It is imperative to avert putting of fly ash generated from power stations on land and encourage fly ash utilisation for construction and other activities.

2 Environment Concerns of Thermal Power Stations

A. *Air Pollution*: The combustion of coal in power plants leads to the formation of air pollutants like particulate matters, SO_2, NO_x, and CO. When these pollutants are emitted to the atmosphere directly, they shall lead to harmful effects on human beings, animals, biota, and buildings. Uncontrolled SO_2 and NO_x emissions from a power plant may also result in problems like acid rain.
B. *Wastewater Discharge*: The highest amount of wastewater generated in a power plant is from a cooling tower. This cooling water blowdown can be recycled after treatment or discharged to a nearby river or sea. When discharged to a water body, it affects the chemical quality of the water body and the resulting thermal pollution can affect the natural temperature of the water body affecting the aquatic flora and fauna. Other wastewater streams from the power plant can be from ash handling, boiler feed water, flue gas desulphurisation, and from other applications in coal-fired power plants.
C. *Ash Handling and Disposal*: Dumping of fly ash in landfills can lead to various environmental problems like soil pollution, land pollution, groundwater pollution, and air pollution in the surroundings. Disposal of fly ash by discarding in badly designed and poorly maintained ash ponds leads to problems like loss of valuable land that can be used for agriculture or farming and also deteriorate the quality of and groundwater surrounding these ash ponds. If the ash ponds are left open without any cover of soil or vegetation, when the fly ash gets dried up, fugitive dust rises from the ponds and contaminates the nearby air raising the overall concentration of particulate matter. Once-through slurry disposal systems place. The disposal of fly ash in slurry form shall further lead to stress on already strained freshwater resources.
D. *Land Degradation*: Keeping in mind the zoning and development criteria, the citation of power stations and other industries is usually done away from the

cities to avoid problems related to resettling or rehabilitation. However, the air pollutant emissions from the stacks are washed out on the downstream side resulting in harmful effects on adjacent population, animals, buildings, and other biotas. The requirement of land for ash ponds needs to be worked out as this land shall be of no use in the future. Also, the hazardous particles from fly ash shall pollute the soil and groundwater in the vicinity of ash pond rendering large chunks of land as non-usable.

E. *Noise Pollution*: The generation of power in power plants shall employ the use of large types of machinery like boilers, belt conveyors, compressors, crushers, fans, pumps, and turbines. All these types of equipment induce a large amount of noise in the nearby areas of their working. Numerous measures installing of silencers for fans and compressors, utilising noise absorbing materials, provision of noise barriers in front of noise-producing areas can be utilised to control the effect of noise generated in power plants. To reduce the occupational exposure of workers, they shall be provided with safety gears for noise reduction and their working hours shall be limited. The noise levels in the surrounding areas can be substantially reduced by providing a green belt surrounding the power plant.

F. *Problems Linked with Fly Ash*: Fly ash generated from the combustion of coal is a fine powder and being lighter travels with air. If fly ash is not handled properly, it can lead to several air pollution problems. Fly ash can reduce the yield of crops if it settles on the soil or vegetation. Disposal of bottom ash and fly ash is done by converting it to slurry from and dumping in ash ponds in the neighbourhood of power plants. Dumping of ash in ponds can lead to various problems in the long-term method. Ash ponds shall need a large chunk of land. This shall reduce the total land available for cultivation or agriculture. Once an ash pond has reached its final height, it is to be covered with soil layer and left unused. Then another land is to be utilised for preparing a new ash pond, leading to additional loss of cultivated land. Large quantities of freshwater shall be utilised for converting fly ash to slurry form further adding to the water requirement. Various salts and metals present in the fly ash slurry shall percolate into the groundwater/soil and pollute it.

3 Environment Protection Guidelines for Ash Disposal

One of the important by-products of coal combustion in power plants is fly ash. Fly ash is generally classified into two types. The first type collected in the bottom of the boiler unit is known as bottom ash and the second type that is entrapped in electrostatic precipitators is known as pulverised fuel ash. Pulverised ash contributes to around 80% of the total fly ash, and the residual 20% is bottom ash. As per the MoEF notification dated 3 November 2009, all thermal power stations that were already in operation shall achieve 100% fly ash utilisation within five years that

Table 1 Fly ash utilisation for thermal power stations that came in operation after 3 November 2009

S. No.	Fly ash utilisation	Time period
1	Minimum 50% fly ash generated	One year from Commissioning
2	Minimum 70% fly ash generated	Two years from Commissioning
3	90% fly ash generated	Three years from Commissioning
4	100% fly ash generated	Four years from Commissioning

is by 3 November 2014. All the thermal power units which came into operation post-notification shall achieve fly ash utilisation as specified in Table 1.

The fly ash that remains unutilised shall be used in two years and the balance unutilised fly ash collected in the first four years shall be used over the next five years. The facilities for utilisation of ash and disposal are to be planned at the initial stage. The factors affecting the environment are fugitive dust emissions, pollution of groundwater and surface water, utilisation of land for ash disposal, and failure of ash dyke.

4 Ash Disposal

Ash is discharged in wet form or dry form. Dry fly ash is generally preferred when used for cement manufacturing.

Dry Ash Disposal: Dry extraction system is adapted for handling and management of ash in dry form. In this system, ash is collected in dry form in the hoppers of electrostatic precipitator; it is then disposed of using a vacuum or a pressure pump. The dry ash is then carried to an additional hopper situated next to electrostatic precipitator. From this additional hopper, by pneumatic conveying, ash is conveyed to a storage site for necessary utilisation. To avoid erosion at the storage site, soil layer is used. As water is absent in ash, generation of the leachate is minimal. Plantation can be done to control fugitive emissions.

Wet Disposal: In this method, the slurry is formed by mixing fly ash with water and is conveyed to landfills for disposal. There are two methods of disposal in a wet system, LCSD which is lean concentration slurry disposal and HCSD which is high concentration slurry disposal. LCSD is conventionally used for disposal of ash slurry in dilute form. However, LCSD has various limitations like a higher amount of water wasted, groundwater pollution, the threat of ash pond collapse, vast land required for ash dyke, higher power consumption, and higher cost for ash pond construction. Looking at these drawbacks, HCSD is adopted which a new environmental-friendly ash disposal policy wherein the higher concentration of ash slurry is used and handled.

5 Collection of Fly Ash

After, coal combustion, 20% of the total ash is bottom ash and 80% is conveyed with stack gases as pulverised fly ash. Bottom ash is converted to slurry after mixing with water or handled in dry form. Generally, electrostatic precipitator is used as an air pollution control device at the stack for collecting fly ash in the flue gases. Usually, dust can be removed from flue gases by using mechanical or electrical methods. For mechanical collection, fabric filters are used and for the electrical method, electrostatic precipitators are used. An ESP is an efficient tool for the collection of dust ranging from sub-micron to larger sizes. An ESP can provide efficiency as close to 99.9%. Operating costs for electrostatic precipitators are low, and they are economically viable in the long run.

6 Characteristics and Composition of Fly Ash

ASTM C618 defines two classes of fly ash, namely Class F fly ash and Class C fly ash. The chemical composition of the ash is the basic difference between Class C and Class F fly. Class F fly ash is known to be highly pozzolanic. When used in cement manufacturing, it undergoes a chemical reaction with excess lime generated in the hydration of Portland cement. Class C fly ash is less pozzolanic and can be self-cementing. The basic physical properties of fly ash are almost similar to that of natural soils; however, they shall vary based on the kind of coal used for power generation. The properties for a typical fly ash sample are shown in Table 2.

Table 2 Properties of fly ash

Property	Typical value
Colour	Whitish grey
Specific gravity	1.9–2.55
Average particle size (μm)	10–100
Bulk density (g/cm^3)	0.90–1.66
Moisture (%)	18–38
Compression index	0.05–0.4
Permeability (cm/s)	10^3–10^5
Angle of internal friction (°)	30–40
Coefficient of uniformity	3.1–10.7
Cohesion	Negligible
Plasticity	Nil

7 Fly Ash Mitigation Measures

The major problems linked with fly ash include the requirement of large amounts of land, water, and energy for its handling and disposal. Again if handling and disposal are not managed properly, the fine particles due to their lightweight may become airborne and add to air pollution woes. Presently, millions of tonnes of fly ash are produced in India annually, with ash ponds occupying large acres of land. Handling such large volumes of fly ash can lead to perplexing problems, for land availability, health risks, and environmental impacts on natural resources. In order to protect the humans, flora, fauna, and environment as a whole the handling, disposal, and reuse of fly ash have to be done with utmost care. Fly ash is a very fine powder, having spherical shape and size ranging from 0.5 to 100 μm. The fly ash particles being very small are capable of reaching pulmonic lungs and staying there for a longer period. Further, these particles can penetrate more deep in the lungs being dumped on the alveolar walls and finally be conveyed to the blood plasma. Fly ash is considered to be non-hazardous referring to the Hazardous Waste Management Rules.

Fly ash generated from the combustion of coal is an exceptional and imminent raw material for producing various materials like fly ash concrete, fly ash bricks, fly ash tiles, and blocks. Fly ash can also be used as a beneficial option in road pavements, embankments, and filling in mining areas which require large volumes of material. Fly ash materials have several noteworthy advantages when compared with regular construction materials. There are several products which can be produced by adding fly ash and gypsum. The probable areas where fly ash is being used currently are:

- Fly ash bricks;
- Fly ash for cement manufacturing;
- Fly ash ceramics;
- Fly ash lightweight aggregate;
- Fly ash polymer products;
- Fly ash as fertiliser;
- Fly ash in road construction;
- Aerated autoclaved concrete (AAC) blocks.

7.1 Fly Ash Bricks

Fly ash is now widely utilised as a partial substitute of clay in the manufacturing of bricks. Fly ash bricks are an eco-friendly option compared to clay bricks as they avoid topsoil erosion, they do not need firing of fossil fuels hence have at least 50–60% lower carbon footprint in manufacturing itself. Since the density of fly ash is half that of the clay, the bricks manufactured from fly ash are light in weight as compared to conventional bricks. Thus, the construction time and dead load are lower. Fly ash bricks help in saving other carbon-intensive construction material like steel. Fly ash bricks have 15–20% lower water absorbability—so they result in 20% lower rendering, and hence, less is mortar needed and water consumption is also less. These bricks have 20% better thermal conductivity, so it shall result in air-conditioning load saving. The production of regular bricks requires a larger quantity of clay as compared to fly ash bricks. Therefore, fly ash bricks help in preserving the topsoil as they need less amount of clay. They also help in reutilisation of fly ash in a constructive way.

Public awareness is required to be generated, and Government shall offer distinct incentives for fly ash utilisation. Orissa Government has prohibited the use of soil for of brick manufacturing within 20 km location of a thermal power unit which has led to the impetus of use of fly ash in brick manufacturing. In the manufacturing of fly ash bricks, clay and fly ash are mixed and burned in the furnace wherein the unburnt carbon in fly ash assists as a fuel in the burning process. Around 20–30% reduction in energy can be achieved by the addition of around 25–40% of fly ash.

7.2 *Fly Ash for Cement Manufacturing*

Fly ash undergoes a chemical reaction with calcium hydroxide in the presence of moisture and carbon dioxide existing in the environment and strikes on the free lime leading to concrete deterioration. It has been detected that challenging free lime is converted to durable concrete with the help of reactive elements existing in fly ash. About 66% of cement used in concrete can be substituted with fly ash while constructing dams and large infrastructure. The substitution of fly ash with cement while using reinforced cement concrete not only results in saving quantity and cement price but also improves the strength and durability. Fly ash when used with Portland cement also improves its performance. It is estimated that manufacturing one tonne of Portland cement releases about 0.87 tonnes of CO_2 emissions. Hence, the substitution of fly ash for cement, in concrete reduces the CO_2 emissions equivalent to the quantity of fly ash substituted by cement, thus reducing the overall carbon dioxide and greenhouse gas emissions.

7.3 *Fly Ash Ceramics*

Low-cost ceramic tiles are produced by utilising fly ash or blast furnace slag. In the production of ceramic tiles, the possibility of using pyrophyllite or mullite to the tune of 50% and fledspar about 16% by weight along with using fly ash from thermal power plants and clays from coal mining has been carried out. The basic properties of fly ash and ceramic mixtures are examined by finding fluctuations in mineralogy and elementary ceramic properties like bulk density, colour, shrinkage, and water absorption. The changes due to heating observed in fly ash and the ceramic bodies are examined by conducting firing tests for temperatures ranging between 900 and

1200 °C. The subsequent ceramic forms display properties, which recommend the use of fly ash in the manufacture of paving stoneware manufacture and for tiling. The fly ash can productively be substituted to the percentage weight between 15 and 50 in diverse kind of ceramic products.

7.4 Fly Ash Lightweight Aggregate

Converting fly ash to a lightweight aggregate presents a principally economic and environmental friendly option for bulk disposal of the large volume of fly ash. The bulk density of fly ash lightweight aggregate falls between 650 and 750 kg/m^3. The manufacture of fly ash lightweight aggregate is extremely appropriate for use in the manufacturing of lightweight concrete. This lightweight concrete produced from lightweight aggregate can be utilised for load-bearing as well as non-load-bearing members. Lightweight concrete produced from fly ash lightweight aggregate has decent potential at places where fly ash is easily obtainable in the vicinity. Fly ash lightweight aggregate has a tremendous perspective of replacing ecologically degrading stone aggregates and riverbed sand, which result in irreparable damage to the environment.

7.5 Fly Ash Polymer Products

Fly ash is also used as a substitute for wood as a construction material. Fly ash is used as the matrix, and jute cloth is placed as the reinforcement. The polymer matrix is formed by passing and coating the jute cloth through fly ash and then curing it. To obtain the required thickness, the number of laminates is augmented. Fly ash polymer matrix can be utilised for several purposes like a door or window shutters, wall panels, and partition panels, ceiling panels or floor tiles. Fly ash polymer as a product is strong, durable, corrosion-resistant, and cost-effective when compared with wood as construction material.

7.6 Fly Ash as Fertiliser

Fly ash enhances the accumulation of chief nutrients like calcium, magnesium, iron, and zinc by crops and vegetation. Therefore, fly ash is deliberated as a probable growth enhancer in soils. Applying fly ash in different doses to the soil can lead to an improvement in the yield of the crop and also improve the water retention capacity of the soil. The application of varying does of fly ash to the soil can increase the yield by about 20–30%. Hence, fly ash serves to be a good soil modifier and fertiliser.

7.7 Fly Ash in Road Construction

Fly ash due to its lightweight and filler properties can be preferably used in making the base surface of roads along with other materials. Since fly ash is available in large quantities and is a waste product from power plants, its use in large quantities for constructing the base and other sources in roads shall end in low-value and high-volume consumption technology. Various projects at Dadri and New Delhi have used fly ash as a filler material in roads or flyover embankments as per the guidelines laid by Indian Roads Congress (IRC). CRRI has recently extended its consultancy services and monitored the construction work near Raichur (Karnataka) for a rural road having length 1 km using fly ash-based flexible and semi-rigid pavement. Another construction area that possesses the substantial capacity for utilisation of fly ash in very large volumes is for embankments in road and flyover. Embankments for flyovers at Okhla, Delhi, and roads at Raichur and Dadri have used very large quantities of fly ash as filler material. It has been estimated if fly ash is used in partial substitution of soil for embankment in roads about Rs. 50–75 per MT of earthwork cost is saved, principally due to saving in excavation and transportation costs.

7.8 Aerated Autoclaved Concrete (AAC) Blocks

Aerated concrete is a sort of light, porous new building material, featured with lightweight, good thermal insulation, easily processing and incombustible, etc. It can be used for making building blocks, plates, and insulation products of different sizes. It is widely used as the bricks of the load-bearing walls or enclosure wall in the industrial and civil buildings. In place of bricks and conventional concrete blocks, lightweight concrete blocks can be utilised as a preferable substitute. It is a German-based technology wherein fly ash is used in the ratio of one-third to one-fourth of the total materials used. Other materials include sand, water, and foam produced from biodegradable foaming materials. The Government of India grants special concession in import duty if more than 25% fly ash is utilised in the manufacture of cellular lightweight concrete blocks using a foaming agent and foam generator. These lightweight concrete blocks are more strong and durable, and they also result in dead load reduction leading to substantial saving in steel and cement costs as well as foundation size. Cellular lightweight concrete blocks possess superior acoustics and thermal insulation properties which further lead to reduced air-conditioning requirements.

8 Environmental Impact of Fly Ash Utilisation

Reutilisation of fly ash reduces the fly ash disposal problem and also aids in using the land resource in a better way. The use of fly ash in road embankments involves covering of fly ash in earthen or RCC panels. This will prevent the seepage of rainwater into the fly ash core during monsoons and thus prevent leach of heavy metals. Fly ash when substituted with cement in concrete undergoes a chemical reaction with cement and further decreases the leaching effect. Fly ash when utilised for stabilisation work undergoes a chemical reaction that further binds the fly ash particles. Therefore, the possibility of occurrence of any form of pollution due to fly ash utilisation in roadworks is insignificant. The developing countries around the world are struggling to overcome various problems like energy deficit, housing for all and rising infrastructure demand. Therefore, the materials preferred for building and infrastructure projects in such developing counties shall be those having low energy demand and hence turning out to be energy efficient. Fly ash produced as a by-product during the combustion of coal if put to proper reuse can be utilised for the manufacturing of fly ash bricks, manufacturing of fly ash concrete, developing fly ash ceramic products, producing fly ash fertiliser, and using fly ash as filler material in road construction and road embankment. This shall reduce a lot of stress on already strained natural resources like land, water, clay and also help in substantial reduction of CO_2 emissions.

9 Discussion

For the production of electricity in India, coal is chiefly utilised as the source of energy in thermal power plants. However, the coal existing in India is of inferior quality, containing a large amount of fly ash in addition to low calorific value. These properties of coal shall immensely affect the properties of fly ash that is produced as a waste by-product from thermal power units. The thermal power plants around the country use a very large amount of coal which results in numerous environmental impacts like stress on water and land resources, global climate change and fly ash handling and management. All over the world, coal-fired thermal power plants are considered to be one of the chief sources of pollution that affect the surrounding environment in terms of air, soil, and water pollution, large land use, health hazards and thus lead to enormous environmental impacts which cannot be reversed. Therefore in order to protect the environment, a proper fly ash handling and management plan are required to be framed to ensure proper disposal of fly ash from the thermal power units. This plan shall consider all possible applications for reutilisation of fly ash. The utilisation of fly ash for brick manufacturing, cement manufacturing, ceramics production, etc., shall be linked with incentives to promote it. Various government and non-government bodies functioning in the field of fly ash utilisation shall promote the utilisation of fly ash for various purposes.

Bibliography

Jha, C. N., & Prasad, J. K. (2000). Fly ash: A resource material for innovative building material—Indian perspective substitute and paint from coal ash. In *2nd International Conference on "Fly Ash Disposal & Utilization"*, New Delhi, India.

Kant, K., & Chakarvarti, S. K. (2003). Environmental impact of coal utilization in a thermal power plant. *Journal of Punjab Academy of Forensic Medicine & Toxicology, 3*, 15–18.

Mishra, U. C. (2004). Environmental impact of coal industry and thermal power plants in India. *Journal of Environmental Radioactivity, 72*(1–2), 35–40.

Senapati, M. R. (2011). Fly ash from thermal power plants–waste management and overview current science. *Current Science (Bangalore), 100*, 1791–1794.

Shikha, S., & Sushma, D. (2011). Effect of fly ash pollution on fish scales. *Research Journal of Chemical Sciences, 1*(9), 24–28.

Vimal, K., & Mukesh, M. (2003). Clean environment through fly ash utilization. In *Cleaner Technology, Impacts/12/2003-2004, MOEF-CPCB*, Govt. of India (235–255).

Scope of Fly Ash Application as a Replacement for Chemical Pesticides for Pest Control in Certain Crop Pockets of Neyveli and Virudhachalam Regions in Tamil Nadu, India

C. Kathirvelu, Y. Hariprasad and P. Narayanasamy

Abstract In an attempt to probe for alternate techniques in place of chemical pesticides consumed in various crop fields and their retention as residues in agricultural soils and various water resources used in irrigation, soil and water samples were gathered from two locations, Neyveli and Virudhachalam adjacent to Neyveli Lignite Corporation India (NLCIL), Neyveli, Tamil Nadu, India, during off-season. Samples were drawn from the fields cropped with rice, sugarcane, cashew nut, mango, groundnut, gingelly intensively grown in this tract, and water samples were gathered from borewells, irrigation canals, lignite mine water, open wells and natural water and analysed for the residues of pesticides. Results revealed that the soil samples drawn from the cropped fields contained various molecules of different groups of pesticides particularly Quinalphos, Endosulfan, Deltamethrin and Etrimphos, whereas the water samples hardly had pesticide molecules. The findings are interpreted to the role of fly ash as dust insecticide found effective from our laboratory against various pests of crops shows promise and appropriateness in the context of replacing chemical pesticides use in agricultural fields. Various chemical pesticides especially of the dust, granule and wettable powder formulations could be dispensed within the event of the fly ash used as a carrier. As a whole, it is inferred that fly ash use in pest control in different crops in agriculture has immense value in the context of protecting the environment and agricultural produces from the chemical pesticides.

Keywords Adjuvant · Silicon · Resistance · Residues · Formulations

---

C. Kathirvelu (✉) · Y. Hariprasad · P. Narayanasamy
Department of Entomology, Faculty of Agriculture, Annamalai University, Annamalainagar, Tamil Nadu 608 002, India
e-mail: ckathirveluau@gmail.com

Y. Hariprasad
e-mail: yagamurhari@yahoo.com

P. Narayanasamy
e-mail: drpnsamy@yahoo.com

S. K. Ghosh and V. Kumar (eds.), *Circular Economy and Fly Ash Management*,
https://doi.org/10.1007/978-981-15-0014-5_2

1 Introduction

Consumption of the chemical pesticides in agriculture, horticulture and stored produces has resulted in the deposition of pesticide residues in the agricultural produces and environment as well as besides being toxic to all the living organisms. Consequently, an array of biological pesticides and varieties of eco-friendly tactics of pest control have been adopted, which have enabled us to keep the chemical pesticides off from the target sites considerably. This has created an awareness among the people to turn towards alternate methods. This chaotic and suffocating situation has triggered us to go in for arriving at alternate techniques of crop protection which are eco-friendly and supportive towards organic agriculture. Therefore, fly ash a coal/lignite-based thermal generated waste appears as an alternate source for the dust, tablet and granular chemical insecticides.

Fly ash is a fine powder and a by-product obtained by burning the coal/lignite in thermal power stations, and it contains amorphous ferro-alumino silicate, similar to soil except organic carbon and nitrogen; thus, it has been found to have a great potential with manifold advantages in agriculture. It has been used in agriculture for its liming nature and also for the essential nutrients which actually enhances the plant growth and also useful to reclaim the nutrient deficiency in soils. Despite the utilization of fly ash in various sectors such as cement, concrete blocks, bricks, etc., still large quantities of the fly ash are heaped on open lands occupying space to make air pollution. This surplus of fly ash could have been shown to be a carrier material in the production of the chemical insecticides. But, fly ash itself has pesticidal value by its constitution of silica and as a hydrophilic compound which could desiccate the insect body and kill them and also wears out the mandibles of insect pests with chewing type of feeding organs. Also, the fly ash owing to its easy availability and cheapness appears to be a gift to perform organic pest control in crops.

2 Fly ash in Pest Control

A maiden attempt was, therefore, made in India by Prof. P. Narayanasamy at the Department of Entomology, Annamalai University, India, as early as in 1988–89 with fly ash (both coal fly ash and lignite fly ash), generated at the Coal/Lignite-based Thermal Power Plants like Neyveli Lignite Corporation India Limited, Neyveli, Tamil Nadu, and elsewhere. The investigation has led to the detection of insecticidal nature of the fly ash as a dust insecticide and an adjuvant/carrier in the formulations of insecticides like dust, granules and wettable powder. Since then, extensive researches were undertaken at this department to reveal to the world confirming the pesticidal and carrier value of the fly ash to anchor in the pest control strategies.

3 Use of Chemical Pesticides in Agriculture in India

As per Food and Agriculture Organization (FAO), Rome, the term "pesticide" means a substance or mixture of substances used to prevent, destroy, or control any pests including human or animal vectors, weed plants, or animals causing annoyance or harm during the production of food and agricultural commodities and their processing, storage, transport, or marketing. The term "pesticide" also includes substances that are intended for use as growth regulator for plants, defoliant against insect pests and also act as desiccant and thinning agent and preventing fall of fruits in premature stage. Also, they are used as substances in crops either before or after harvest to protect the commodity from deterioration during storage and transport.

There is a wide regional variation in the use of pesticides across various states in India. The state-wise consumption of chemical pesticides during 2016–17 in India had indicated that Maharashtra, Uttar Pradesh, Punjab and Haryana consume maximum quantity of pesticides in India. Maharashtra is the leading consumer with 13,496 metric tones annually whilst Punjab, Haryana and Telangana were followed suit with 5483, 4050 and 3840 metric tonnes, respectively (DPPQS Report 2018).

Though crop losses due to pest attack in India are reported to be very high, the intensity of pesticide consumption in India is one of the lowest when compared to other countries in the world. The pesticide usage in India is only about 9% of the total cultivated land (16.7 million hectares). The state, Jammu and Kashmir, ranks first in the intensity of pesticides application with an average level of 2.337 kg/ha, followed by Punjab (1.377 kg/ha) and Haryana (1.151 kg/ha). North-eastern states (Sikkim, Nagaland, Assam, Manipur, Arunachal Pradesh), Rajasthan and Madhya Pradesh were handled pesticides at less than 100 g/ha of cropped area (Indira Devi et al. 2017). In India, cotton, paddy, vegetables and fruits are grown in 32% of the cultivated area and account for about 80% of the pesticides consumption (Table 1).

Table 1 Crop-wise consumption of pesticides in India

Crops	Share of pesticide use (%)
Cotton	44.5
Paddy	22.8
Jowar	8.9
Fruits and vegetables	7.0
Wheat	6.4
Arhar	2.8
Others	7.6
Total	100

*As per Annual Report 2009, Department of Chemicals and Petro Chemicals, Government of India

The State Punjab tops the list in per hectare consumption of pesticides as 0.74 kg, followed by Haryana (0.62 kg) and Maharashtra (0.57 kg) during the year 2016–17, while Bihar, Rajasthan, Karnataka and Madhya Pradesh consumed pesticides at lower level.

4 Detection of Insecticide Residues in Soil and Irrigation Water in the Study Area

Pesticide residues in the process of crops/vegetables/fruits pose danger to human health. There are possibilities of entry of pesticides in the food chain posing danger of bio-magnification. Vegetables like cabbage, cauliflower, brinjal, tomato, okra, capsicum, cucumber, bitter guard, green pea; fruits like apple, banana, guava, grape, pomegranate, mangoes, orange; spices like cumin, pepper; red chilli powder, rice, wheat, pulses, fish, meat and tea and surface water contained residues of organochlorine, organophosphorus, synthetic pyrethroids, carbamates and herbicides (Lok Sabha Standing Committee Report 2016).

An attempt was made to probe consumption of chemical pesticides in various crop fields and their retention as residues in the agricultural soils and various water resources used in crop irrigation. Series of surveys on soil and water sample were carried out in two tracts, namely Neyveli and Virudhachalam (Fig. 1) areas adjacent to Neyveli Lignite Corporation India Limited (NLCIL), Neyveli, lying 45 km away north of Annamalai Nagar in Cuddalore District of Tamil Nadu and the samples gathered during off-season. Samples were drawn from the fields cropped with rice, sugarcane, cashew nut, mango, groundnut, gingelly and also water samples gathered from borewells, irrigation canals, lignite mine water, open wells and natural water.

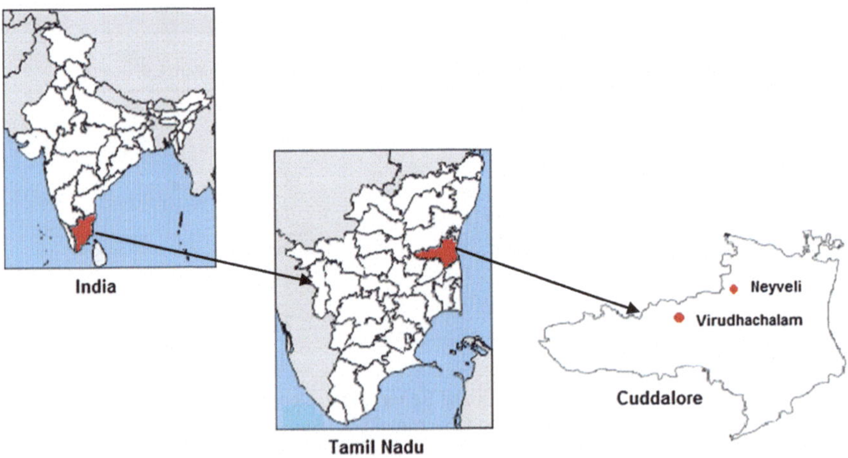

Fig. 1 Study area map

The soil and water samples collected so were analysed for the residues of pesticides using GCMS.

Analyses of soil samples revealed the presence of insecticides such as DDT (o,p-DDE, o,p-DDD, o,p-DDT, p,p′-DDE, p,p′-DDD, p,p′-DDT), lindane (r-HCH), BHC (α-BHC-HCH, β-BHC-HCH, γ-BHC-HCH), endosulfan (α-endosulfan, endosulfan, endosulfan sulphate) monocrotophos, ethion, chlorphyrifos, phorate (phorate, phorate sulphon, phorate sulphoxide), methyl parathion (methyl parathion, methyl paraoxon), malathion (malathion, malaoxon), aldrin (aldrin, dieldrin) diazinone, dichlorvos, etrimphos, chlorpyrifos—methyl, metalaxyl, heptachlor, fenitrothion, triadimifon, chlorfenvinfos, quinalphos, cis-chlrodane, profenofos, triazophos, benalaxyl, phosalone, azinphos-methyl, permethrin, cypermethrin, cyfluthrin, fenvalerate and deltamethrin (Table 2) (Coal India Project Report 2011).

Except the crops like mango, soil samples drawn from the land grown previously with rice, black gram, sugarcane, groundnut and cashew nut were found with pesticide molecules, whereas the water sources had no pesticide residues (Table 3). Similarly all the water samples from the borewell, irrigation channel effluent river and open well did not have pesticides. As the area is continuously cropped one, the presence of pesticide residues in the respective cropped soil might be due to continued application of chemical insecticides as dusts, emulsifiable concentrate and wettable powder formulations. The practice of continuous application of pesticides for every crop, there is a chance of accumulation of toxic element in the soil.

In support of this finding, Lari et al. (2014) observed that there was high concentration of organochlorine (endosulphan, HCH, aldrine, dicofol, DDT, etc.) and organophosphates (dichlorvos, ethion, parathion, methyl phorate, chlorpyriphos, profenofos) found in groundwater compared to surface water. Of them, organophosphates were found more contaminated than the organochlorines.

Traces of pesticides could be detected in all the spheres of environment, air, water and soil. Yadhav et al. (2015) reported that consumption of persistent organic pesticides in the environment samples from India, residue level was high in the water and the soil and asserted that India is one of the major contributors of persistent organic pesticides consumption and no significant decline in DDT and HCH levels was observed despite their ban.

In the second phase, the soil samples drawn from the lands cropped with cashew nut, groundnut and lime trees in places like Periyankuppam, Muthandikuppam, Seduthankuppam and Clonal seed orchard lying around Neyveli contained molecules of certain pesticides (Table 4).

In the second phase of water samples drawn from the water resources like borewell and lake in and around the NLCIL, Neyveli area had no trace of pesticides (Table 5).

In the light of the above revelations, it is inferred that by the practice of the intensive agriculture adopted in the Neyveli and Virudhachalam tracts with the application of various kinds of pesticides in various crops, pesticides of organophosphatic and organochlorinated categories had got accumulated and settled in the soils and waters analysed. This is highly detrimental for the feature of contamination of foodstuffs posing health hazard to the people of the region.

Table 2 Analyses of soil samples for levels of insecticide residues in Neyveli and Virudhachalam areas of Cuddalore District, Tamil Nadu, during 2008 (I Series)

Soil sample No.	Field location	Crop	Quantum of pesticide residues (μg/kg of soil) Detection level (0.01 μg/kg)
1	Vayalur	Rice fallow Black gram	Nil
2	Vayalur	Fallow land	Nil
3	Kothavachery	Rice fallow Black gram	α-HCH—0.04 β-HCH—0.08 γ-HCH—0.05 diazinone—0.08 etrimphos—0.10 chlorpyrifos-methyl—0.03 fenitrothion—0.03 chlorpyrifos—0.03 aldrin—0.06 parathion—ethyl—0.04 triadimifon—0.03 quinalphos—0.06 o,p′-DDE—0.09 p,p′-DDE—0.03 dieldrin—0.06 o,p′-DDD—0.07 endosulfan- β—0.10 ethion—0.03 p,p′-DDD—0.08 o,p′-DDT—0.02 permethrin—1—0.01 permethrin—2—0.03
4	Thermal Power Station II (NLCIL) (Andarkollai)	Sugarcane	Nil
5	Thermal Power Station II (NLCIL) (Andarkollai)	Groundnut	Permethrin—1—0.05 Deltamethrin—1—0.16
6	Romapuri (Mandarakuppam)	Sugarcane	α-HCH—0.04 β-HCH—0.08 γ-HCH—0.03 diazinone—0.04 etrimphos—0.07 chlorpyrifos-methyl—0.02 fenitrothion—0.02 parathion—ethyl—0.03 quinalphos—0.04 o,p′-DDE—0.04 p,p′-DDE—0.01 o,p′-DDD—0.02 ethion—0.01 p,p′-DDD—0.03 Cypermethrin—2—0.06

(continued)

Table 2 (continued)

Soil sample No.	Field location	Crop	Quantum of pesticide residues (μg/kg of soil) Detection level (0.01 μg/kg)
7	Sathamangalam (Vridhachalam)	Cashewnut	Endosulfan—β—0.10
8	State Horticultural Farm (Block-4, Neyveli)	Mango	Nil

Table 3 Analyses of water samples for levels of insecticide residues in Neyveli and Virudhachalam areas of Cuddalore District, Tamil Nadu, during 2008 (I Series)

Water sample No.	Location	Actual site of sampling	Source	Amount of pesticide residues (μg/kg) Detection level (0.01 μg/kg)
1	Keerapalayam	Cropped area	Borewell	Nil
2	Sathapadi	Cropped area	Borewell	
3	Thermal Power Station II (NLCIL)	Cropped area	Irrigation Canal flowing in Mines water	
4	Velikunankurichi	Cropped area	Borewell (NLCIL)	
5	Sathamangalam (Vridhachalam)	Cropped area	Borewell	
6	Sathamangalam (Vridhachalam)	Cropped area	Dug well	
7	Mandarakuppam	Mines II-Effluent Water	Effluent river	
8	Romapuri (Mandarakup-pam)	Cropped area	Open well	
9	State Horticultural Farm (Block-4, Neyveli)	Mango Orchard area	Borewell	

Table 4 Analyses of soil samples for levels of insecticide residues in and around Neyveli and Virudhachalam areas of Cuddalore District, Tamil Nadu, during 2008 (II Series)

Sample No.	Field location	Actual site of sampling	Crop grown	Amount of pesticide residues Detection level (0.01 mg/kg)
1	Nursery yard Department of Horticulture, Block-19, NLCIL, Neyveli.	Amidst crop	Coconut garden	Nil
2	Office of Horticulture Division, Sub-Division-I, NLCIL, Airstrip, Neyveli.	Amidst crop	*Plumeria rubra* plantation	Nil
3	Tamil Nadu Forestry Extension Centre, Forest Genetics Division, State Forestry Research Institute, Vridhachalam research range, Block-B, Indranagar	Rainwater recharge Lake	*Eucalyptus globulus* plantation	Nil
4	Clonal Seed Orchard, Tamil Nadu Forestry Research Centre, Tamil Nadu Forestry Division, Block-A, Indranagar	Rainwater recharge Lake	*E. citriodora* plantation	p,p′-DDE—0.01 ppm p,p′-DDT—0.015 ppm
5	Marungur	Amidst crop	Cashewnut plantation	Nil
6	Sorathur	Amidst crop	Cashewnut plantation	Nil

(continued)

Table 4 (continued)

Sample No.	Field location	Actual site of sampling	Crop grown	Amount of pesticide residues Detection level (0.01 mg/kg)
7	Periyankuppam	Amidst crop	Cashewnut plantation	Quinalphos-2.11 ppm Endosulfan sulphate—0.026 ppm
8	Muthandikuppam	–	Fallow land	p,p′-DDE—0.025 ppm p,p′-DDT—0.020 ppm Quinalphos—0.03 ppm
9	Seduthankuppam	Amidst crop	Cashewnut plantation	Nil
10	TANCOFF State Oilseed Farm, Block-30, Neyveli	Amidst crop	Groundnut field	Nil
11	Permanent Fly ash Trial Plot, CARD Complex, Neyveli	Amidst crop	Groundnut field	Nil

5 Comparison of Fly Ash Cost with Chemical Insecticides

The fly ash is being sold at 450 rupees (US$6.5) a ton at the thermal power station of NLCIL, Neyveli for use in agriculture and brick making. Hence, the fly ash could be procured and applied in the field to control various crop pests which involve around Rs. 20 (US$0.3) for 40 kg/ha. It clearly indicates that the use of fly ash would be fivefold to tenfold cheaper than the application of the chemical insecticides in the form of granules, dusts and wettable powders (Table 6).

6 Scenario of Insecticides Residue-Free Agriculture Compared with Fly Ash Application

In order to deliver pesticide-free healthy food and vegetables to the people, there is a mechanism which is in operation in the country through a number of pesticides testing laboratories to take care of the produces being contaminated with the persistent pesticides. Therefore, all the vegetables and fruits produced should be completely

Table 5 Collection of water samples in and around Neyveli and Virudhachalam areas of Cuddalore District, Tamil Nadu, during 2008 (II Series)

Sample No.	Field location	Utility	Source	Amount of pesticide residues (μg/kg) Detection level (0.01 μg/kg)
1	Nursery, Department of Horticulture, Block-19, NLCIL, Neyveli	Irrigation water	Borewell	Nil
2	Office of Horticulture Division, Sub-Division-I, NLCIL, Airstrip, Neyveli	Irrigation water	Borewell,	
3	Tamil Nadu Forestry Extension Centre, State Forestry Research Institute, Vridhachalam research range, Block-B, Indranagar, Neyveli	Irrigation water	Borewell	
4	Clonal Seed Orchard, Tamil Nadu Forestry Research Centre, Block-A, Indranagar, Neyveli	Irrigation water	Borewell	
5	Marungur	Drinking water	Borewell	
6	Sorathur	Natural water	Lake	
7	Periyankuppam	Natural water	Lake	
8	Muthandikuppam	Natural water	Lake	
9	Seduthankuppam	Drinking water	Borewell	
10	TANCOFF State Oilseed Farm, Block-30, Neyveli	Irrigation water	Borewell	
11	Permanent Fly ash Trial Plot, CARD Complex, Neyveli	Irrigation water	Borewell	

free of the pesticides. In support of this, many trained personnels working in various organizations like consultancy centres, advisory cells, farming consultants, farm process consultant, poultry and horticulture consultancy, agri-tech consultancy, rainwater consultancy operate in the country should come forward to meet the challenge and to guide the common man to lead pesticide-free environment and food.

Considering the impact of the pesticide residues in the environment, agricultural produces and animal flesh, etc., an International Workshop 'Seeking agricultural

Table 6 Cost of certain common insecticide formulations marketed in Tamil Nadu, India

S. No.	Name of the insecticide	Formulations	Approximate price/kg/litre (Rs.)	US$
1	Carbofuran 3% G	Granules	95	4.00
2	Phorate 10% G	Granules	80	1.15
3	Fipronil 0.03%G	Granules	90	4.00
4	Thiomethoxam 20% + Chlorontroniliprole20% WG	Granules	250	1.65
5	Cartap Hydrochloride 4% G	Granules	85	1.20
6	Fibronil 40% + Imidacloprid 40%	Wettable granules	1200	17.0
7	Methyl Parathion 2% D	Dust	30	0.43
8	Chlorpyriphos 12% D	Dust	30	0.43
9	Malathion 50% D	Dust	30	0.43
10	Diafenthiuron 50% WP	Wettable powder	300	4.30
11	Acephate 50% + Imidacloprid 1.8 SP %	Soluble powder	1300	18.50
12	Acephate 75%WP	Wettable powder	750	10.70

produce free of pesticide residues' held in Indonesia during 1998 concluded that by minimizing pesticide input, containing the pesticide consumption and selective for use of the pesticides having minimal environmental impact, there is a need to have an effective and inexpensive means of monitoring the possibility of impacts on food production and the environment (Kennedy et al. 1998).

7 Effect of Silica in Crop Protection

Substantial research on the impact of silica showed itself as a functional nutrient for crops like rice, sugarcane. Silica deposition in monocots provides a physical barrier in plants against the insect pest's infestation. Silica was found to enhance the resistance of the plants against pests and diseases significantly with the increase in yield. Hence, the silica plays an active role in terms of physiological resistance factor in crop plants against the pests and the diseases through the production of tannic and phenolic components, among others. The use of silicon in the crop field will be a viable component of integrated management of the insect pests and the diseases and can be integrated with new tactics of pest control (Laing et al. 2006). They concluded that the silica application must include validation for pest control, identification of good sources and their optimum dosages for pest control in various crops.

Silica is usually considered as one of the most important beneficial elements for rice production as the rice plant requires large amounts of silica for its growth. It was estimated that for the production of 100 kg brown rice, about 20 kg of silica is likely to be removed from the soil by the rice plants (Dobermann and Fairhurst 2000). Sharma and Chatterji (1972) found that the application of silica in maize (*Zea mays* L.) provided resistance to the stalk borer (*Chilo zonellus* Swinhoe) damage.

It was detected that fly ash application into the rice root system had accentuated layers of sclerenchymatous cells in the leaf lamina. Such thickened cell layers had caused wearing of the mandibles of rice yellow borer larvae when they attempted to chew the leaves, rendering the larvae unable to feed and died later on due to hunger (Narayanasamy 1997). Similarly, sucking pests like brown planthoppers too infesting rice had intimidating effect of such plants when they tried to taste the plant sap.

Rice husk ash which contains high amount of silica has been effectively used to check the insects and pests in the stored foodstuffs. When it was used as oil spill against *Callosobruchus maculatus* (F) and *Sitophilus zeamais* (Mots), it was found effective when used with waterproofing chemicals, flame retardants and also as a carrier for insecticides and other groups of pesticides (Kumar et al. 2012). The potato tubers are stored for up to 5 months to free from the infestation of potato tuber moth (*Phthorimaea operculella* Zell.) (Das and Rahman 1997). This indicates the inclusion of rice husk ash along with fly ash for pest control operations.

8 Conclusion

It was evident that the application of insecticides for the management of various pest's species of cultivated crops leave behind residues in the soil while water samples had scarcely. This is very much evident from the findings of the study conducted in the aforesaid tracts. It is time for awakening that the application of insecticides must be at minimal dose synchronized with other components of integrated pest management practices, including the application of fly ash as dust and carrier in insecticide formulations for pest control in monocot crops like rice, sugarcane and maize to tackle chewing and sucking pests. It is felt that the findings obtained as aforesaid for the two tracts in Tamil Nadu might auger well for application for other areas practicing intensive agriculture.

Acknowledgements The authors are thankful to the Ministry of Coal, Government of India, for granting with financial support for a four-year research project on fly ash. Similarly, our grateful regards are due to the authorities of Annamalai University, Tamil Nadu, for permitting the project and for providing necessary infrastructure facilities for the successful conduct and completion of the project in time.

References

Coal India Project Report. (2011). *Project completion report on development and use of flyash based pesticides funded by Ministry of Coal*. New Delhi: Government of India.

Das, G. P., & Rahman, M. M. (1997). Effect of some inert materials and insecticides against the tuber moth, *Phthorimaea operculella* in storage. *International Journal of Pest Management, 43*(3), 247–248.

Directorate of Plant Protection, Quarantine and Storage Report. (2018). Pesticide monitoring and documentation. ppqs.gov.in. Accessed on August 29, 2018.

Dobermann, A., & Fairhurst, T. (2000). Rice: Nutritional disorders and nutrient management. IRRI, Philipphines.

Indira Devi, P., Thomas, J., & Raju, R. K. (2017). Pesticide consumption in India: A spatio-temporal analysis. *Agricultural Economics Research Review, 30*(1), 163–172.

Kennedy, I. R., Skerritt, J. H., Johnson, G. I., & Highley, E. (1998). Seeking agricultural produce free of pesticide residues. In *Proceedings of an International Workshop* held in Yogyakarta, Indonesia, 17–19 February, 1998.

Kumar, A., Mohant, K., Kumar, D., & Parkash, O. (2012). Properties and industrial applications of rice husk: A review. *International Journal of Emerging Technology and Advanced Engineering, 10,* 2250–2459.

Laing, M. D., Gatarayiha, M. C., & Adandonon. (2006). A silicon use for pest control in agriculture: A review. *Proceedings of South African Sugar Technology, 80*, 278–286.

Lari, S. Z., Khan, N. A., Gandhi, K. N., Meshram, Tejal S., & Thacker, N. P. (2014). Comparison of pesticide residues in surface water and ground water of agriculture intensive areas. *Journal of Environmental Health Science and Engineering, 12,* 11.

Lok Sabha Standing Committee 29th Report on Agriculture. (2016). Impact of chemical fertilizers and pesticides on agriculture and allied sectors in the country. Ministry of Agriculture and Farmers Welfare (Department of Agricultural Research and Education). Govt. of India. Lok Sabha (Fourteenth Edition), New Delhi.

Narayanasamy, P. (1997). Final report of scheme titled *Studies on use of lignite flyash as an insecticide and an adjuvant in insecticide formulations* supported by the Tamil Nadu State Council for Science and Technology, Govt. of Tamilnadu, Adayar Chennai, 99p.

Sharma, V. K., & Chatterji, S. M. (1972). Studies on the nutritional deficiencies in maize in relation to stem borer, (*Chilo partellus* Swinhoe) resistance. *Indian Journal of Entomology, 34*, 5–10.

Yadav, I. C., Devi, N. L., & Syed, J. H. (2015). Current status of persistent organic pesticides residues in air, water and soil and their possible effect on neighboring countries: a comprehensive review of India. *Science of the Total Environment, 511,* 123–137.

References

Fly Ash and Its Utilization in Indian Agriculture: Constraints and Opportunities

Ch. Srinivasa Rao, C. Subha Lakshmi, Vishal Tripathi, Rama Kant Dubey, Y. Sudha Rani and B. Gangaiah

Abstract Indian electricity generation is majorly dependent on thermal energy by burning the coal producing large amount of fly ash as by-product. Dumping and disposal of fly ash in ponds and land is a routine practice which raises various environmental concerns. Hence, the Ministry of Environment, Forest and Climate Change (MoEFCC), Govt. of India has made continuous efforts for proper utilization and disposal of fly ash. This by-product's rich nutrient content has opened doors for its utilization in agriculture rising a tremendous potential in improving crop productivity and soil health. Besides its nutrient efficiency, fly ash treatment showed a significant result in agricultural insect-pest control. However, agricultural use of fly ash is quite limited in comparison to other sectors of India. MoEFCC has revised norms of fly ash usage and also made a mandate for supplying fly ash free of cost to farmers in the radius of three hundred kilometres. Fly ash is also an excellent substitute for reclamation of low-lying areas and helps in restoration and protection of topsoil layer; with an ever-increasing demand for electricity production in India, fly ash production will also increase. Thus, it is high time to explore the untapped potential of fly ash utilization in Indian agriculture for its sustainable management particularly for timber, ornamental, jute and fibre crops and other agriculture and food systems after proper quality testing.

Keywords Agriculture productivity · Fly ash · Land restoration · Soil health

Ch. Srinivasa Rao (✉) · C. Subha Lakshmi · V. Tripathi · R. K. Dubey
ICAR-National Academy of Agricultural Research Management, Hyderabad 500030, India
e-mail: chsrao_director@naarm.org.in

Y. Sudha Rani
Acharya NG Ranga Agricultural University, Bapatla Guntur District, Andhra Pradesh 522101, India

B. Gangaiah
ICAR-Central Island Agricultural Research Institute, Port Blair, Andaman & Nicobar Islands 744105, India

V. Tripathi · R. K. Dubey
Institute of Environment & Sustainable Development, Banaras Hindu University, Varanasi, Uttar Pradesh 221005, India

S. K. Ghosh and V. Kumar (eds.), *Circular Economy and Fly Ash Management*,
https://doi.org/10.1007/978-981-15-0014-5_3

1 Introduction

Improper combustion of coal/lignite in coal-based power generation leads to fly ash production consisting of partially burned minerals of coal. Fly ash properties differ according to coal type, ash content in coal, boiler type and method of combustion. Large scale production of fly ash causes a problem for its disposal. Fly ash is generally disposed near the power stations as dry ash or pond ash. The fine structure (0.01–100 μm particle size), high surface area, spherical-shaped particles and the lightweight nature of fly ash leads to widespread contamination of the environment originating from its primary disposal site. Radionuclides like thorium and uranium and toxic metals viz. B, V, Cr, As, Hg, Se, Cr make fly ash a serious environmental hazard. Fly ash has the potential to alter soil ecology, water and air causing serious pollution. Moreover, fly ash particles cause corrosion of surfaces and its inhalation causes respiratory problems in human beings. Further, Indian coal has 35–38% ash content which is significantly higher in comparison to the 10–15% ash content of imported coal. Nearly 70% of the power is produced from thermal power stations leading to the production of huge amount of fly ash requiring large land area for its disposal in India. Dumping and disposal of fly ash is highly unsustainable as it creates the threat of source point contamination and raises various environmental concerns. This is the reason that Ministry of Environment, Forest and Climate Change (MoEFCC) has made consistent efforts towards proper utilization and clearance of fly ash. Globally about 800 million tons of coal ash is produced per annum (Costa 2016). India stands next to China and USA in fly ash production. During 2016–17, India generated around 169.25 million ton (Mt) of fly ash of which only 107.09 Mt was utilized. Seven states including Andhra Pradesh, Chhattisgarh, Madhya Pradesh, Maharashtra, Odisha, Uttar Pradesh and West Bengal have contributed to more than 10 Mt of fly ash production. Among these states, Uttar Pradesh has produced maximum fly ash with a value of 28.27 Mt during the year 2016–17 (Fig. 1).

The fly ash production in India has elevated from 68.88 to 169.25 Mt (9.63–63.28%), i.e. around 2.45 times during 1996–97 to 2016–17 (9.63–63.28%) (Fig. 2). Still, 36.72% of fly ash remains unutilized in India posing a potential threat to the environment and human beings.

Despite the fact that fly ash is globally considered as obnoxious waste, the presence of essential nutrients and minerals in fly ash makes it a valuable resource for agricultural amendments to improve plant health and reduce nutrient impoverishment in the soil. The present chapter discusses the characterization and utilization of fly ash in agriculture and the limitations associated with its use.

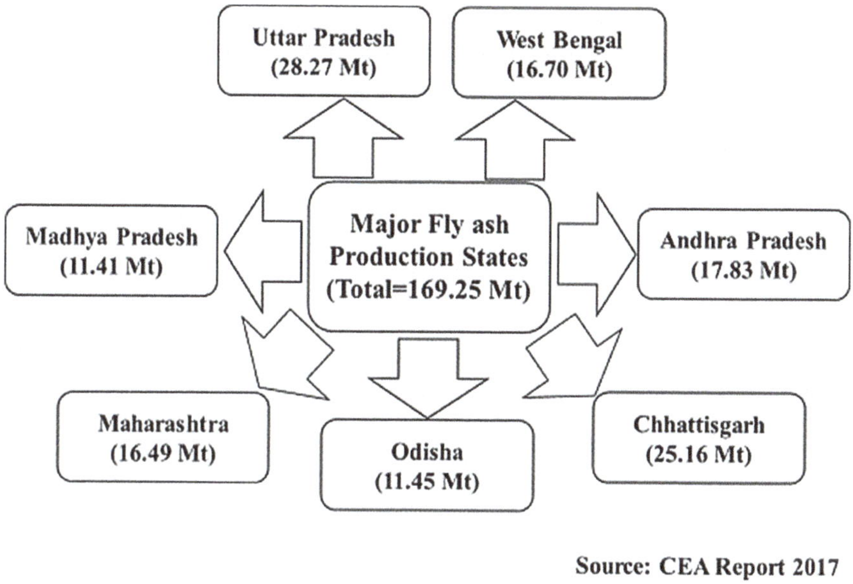

Fig. 1 Major fly ash producing states of India

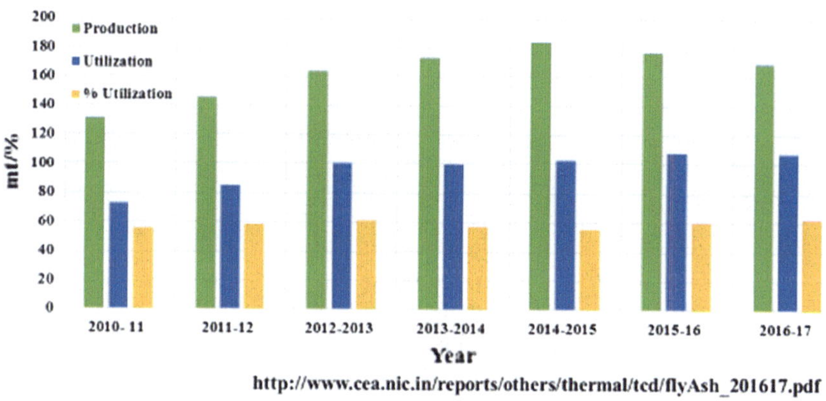

Fig. 2 Fly ash production and utilization in India during the period 1998–99 to 2016–17

2 Characterization of Fly Ash

2.1 Physical, Chemical and Mineralogical Properties

Fly ash's physical properties differ with type of coal, boiler, coal ash content, combustion procedure and collector configuration. It usually has silt loam texture having 65–90% of the particles with diameter of <0.010 mm (Chang et al. 1977; Roy et al.

1981). Particles of Fly ash are empty spheres referred as cenospheres which are air borne, easily and comprises around 1% of the total mass, filled with crystals and minute particles called plerospheres. Low bulk density (1.01–1.43 g/cm^3), specific gravity (1.6–3.1 g/cm^3) and hydraulic conductivity have been reported for fly ash (Roy et al. 1981; Mattigod et al. 1990; Tolle et al. 1982). Further, the average particle densities for non-magnetic and magnetic fly ash particles are 2.7 and 3.4 g/cm^3, respectively, while the moisture retention varying from 6.1% at 15 bars to 13.4% at 1/3 bar (Natusch and Wallace 1974).

Accountable chemical components of fly ash includes oxides of calcium, aluminium, silicon and iron (95–99%); secondary components viz. oxides of K, Mg, Na, P, S, and Ti, (0.5–3.5%) and unburned C (roasting loss), trace elements like As, Ba, B, Cd, Cr, Co, Cu, Mo, Mn, Ni, Pb, Se, Sr, V, Zn, etc. (0.1–0.3%) (Fig. 3) and radioactive elements like ^{40}K, ^{226}Ra, ^{232}Th. Fly ash pH generally ranges between 4.5 and 12.0, depending on the S level in the coal, combustion and sulphur removal technology. The fly ash wastes with Ca/S ratio <2.5 are reportedly acidic in nature and ratio >2.5 makes the waste alkaline (Feng et al. 2006; Szponder and Trybalski 2009; Pandey and Singh 2010).

Fly ash contains mainly amorphous substances like silicon–aluminium glass (80%), few crystalline phases such as gypsum, mullite, quartz, magnetite, anhydrite, ettringete, opaline, haematite, lime, spinel, chlorite and feldspar (Ahmaruzzaman 2010; Szponder and Trybalski 2009). The nutrient content (%) is shown in Fig. 4 and the quality of Indian coal is shown in Table 1.

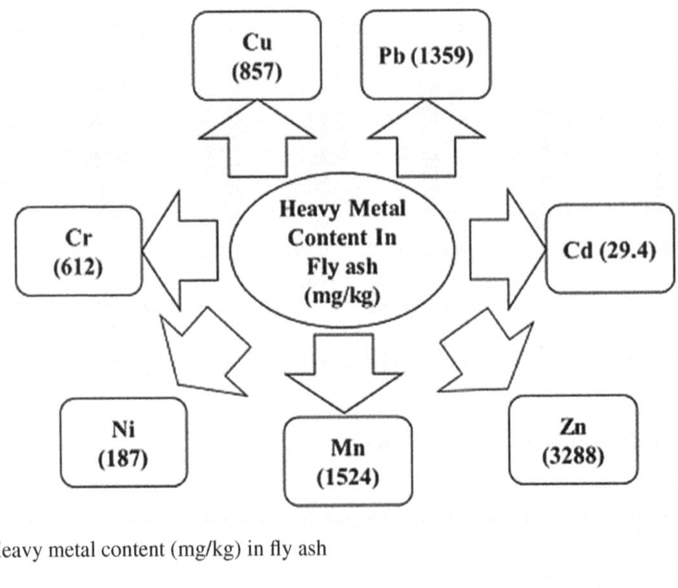

Fig. 3 Heavy metal content (mg/kg) in fly ash

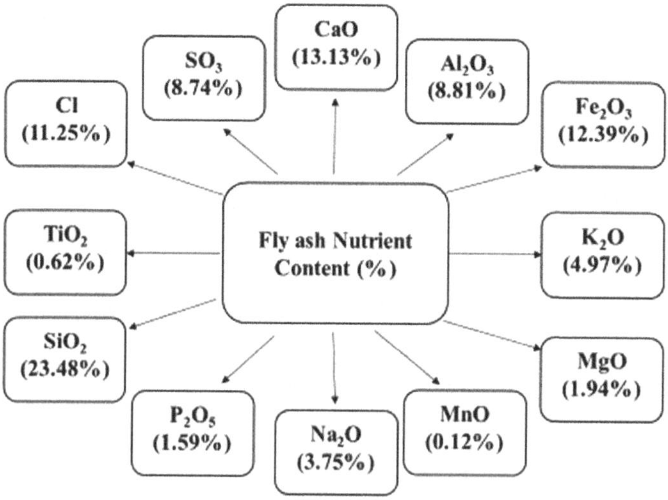

Fig. 4 Nutrient content (%) in fly ash

3 Multiple Uses of Fly Ash

3.1 Fly Ash Utilization Planning (Nearby Thermal Power Plants)

Fly ash being residue of coal power plants available near to the industry is generally used in landfills. In the production of hydraulic cement/plaster and a fully or partial substituting material in Portland cement nearly 43% is recycled, often as pozzolan in concrete production (Sett 2017). By virtue of hydraulic or pozzolanic activity or both fly ash when used in conjunction with Portland cement as a supplementary cementing material, contributes to the properties of hardened concrete. Fly ash is also used in the manufacturing of cement as a filler material, concrete, ceramics, various constructions and as road base admixture such as asphalt mix. Low-lying area fill, dyke raising, roads and embankments and brick manufacturing are other areas of fly ash application. Another viable option could be spreading of fly ash in forest and agricultural lands as a soil stabilizer. Owing to being alkaline and having the presence of various elements, use of fly ash on soil has a remarkable impact on physico-chemical properties of soil from the point of crop production. Percentage utilization of fly ash in different sectors is depicted in Fig. 5.

Table 1 Comparison of fly ash generated from Indian coal and imported coal

Coal field/state	C	H	S	N	O	Ash fraction	Moisture	CV (kcal/kg)	FA fraction
Indian coal quality (per cent by weight)									
Raniganj, West Bengal	60.2	4.2	0.3	1.8	7.1	22.9	3.50	4280	82.0
Giridih, Jharkhand	50.8	3.8	0.5	1.3	5.9	34.0	3.70	4058	83.5
East Bokaro, Jharkhand	61.0	4.10	0.49	1.53	7.2	21.0	4.68	4300	83.50
West Bokaro, Jharkhand	52.30	3.30	0.41	1.23	4.9	36.0	1.86	4098	79.60
Sngrauli, MP and UP	39.27	2.80	0.55	0.92	0.18	39.0	8.28	3850	78.25
Korba, Chhattisgarh	42.93	2.81	0.40	1.06	8.5	38.0	6.30	3997	90.00
Wardha Valley, Maharashtra	46.40	2.9	0.41	1.16	9.3	42.0	7.83	4020	88.58
Imported coal quality (per cent by weight)									
Australian coal	70	4.8	0.5	1.7	6.3	14.4	2.3	6990	70
American coal									
Pennsylvania	77.4	5.2	2.4	1.4	5.0	7.5	1.1	7728	55
Ohio	64.2	5.0	1.8	1.3	8.9	16.0	2.8	6378	51
West Virginia	70.0	5.1	1.2	1.5	12.9	8.1	1.2	7522	58
Kentucky	70.9	5.1	2.3	1.5	8.1	9.8	2.3	7022	60
Illinois	73.7	5.1	2.3	1.6	7.4	7.9	2.0	7330	55.5

Source Chandra and Chandra (2004)

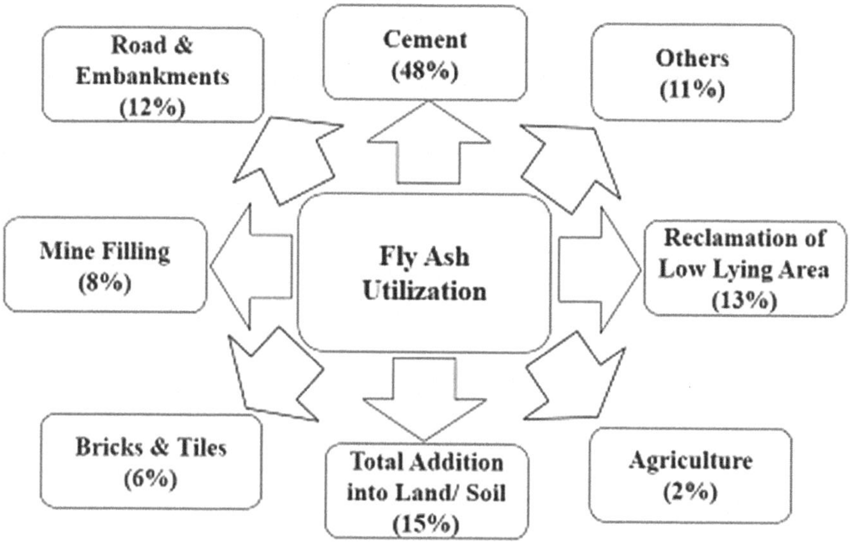

Fig. 5 Fly ash utilization (%) in different sectors

3.2 Agriculture: Introduction to Use Fly Ash

The nutrient and trace element content of the fly ash makes it a valuable resource for soil amendment. Fly ash amendment in low amount is shown to promote plant growth by providing nutrients. Further, the alkaline pH of the fly ash also favours the plant growth. Fly ash has considerable effects for both soil amendment as well as enrichment of nutrients to enhance growth of crops and yield. The utilization of fly ash improved the yield level of many important crops, viz. field corn, wheat, barley, sorghum, green gram, potatoes, onion, carrots, beans, cabbage, tomatoes and alfalfa (Garg et al. 2005; Furr et al. 1978). Application of lower doses (10–25%) of fly ash helped in increasing the growth of rice plants and also increased the non-protein thiol and cysteine levels around 100% in comparison to no fly ash application, resulted in uptake of metals, viz. As, Cd, Cu, Fe, Mn, Ni, Si, Zn, but major of them were confined to root. Interestingly, As accumulation was only reported in leaves and was not detected in roots and seeds (Dwivedi et al. 2007). In lettuce, a 5% application of coal fly ash registered increased seed germination percentage and root depth (Lau and Wong 2001). Oilseed crops such as sesame (*Sesamum indicum*) and groundnut (*Arachis hypogaea*) (Thetwar et al. 2006; Sao et al. 2007; Basu et al. 2007) also exhibited enhanced performance with the application of fly ash. *Brassica juncea* grown with 25% fly ash amendment showed increased plant biomass production (Gupta and Sinha 2006). Different fly ash soil combinations helped to accomplish enhanced yields in aromatic grasses, mainly citronella (*Cymbopogon nardus*) and palmarosa (*Cymbopogon martini*) which can be due to the increased nutrients availability to plant (Asokan et al. 1995; Neelima et al. 1995). Fly ash could also be used for supporting agroforestry as tree species like *Dalbergia sissoo*, *Acacia nilotica*, *Azadirachta indica*, *Eucalyptus hybrid*, *Tectona grandis* and *Tamarindus indica* are known to perform well in fly ash-amended soil (Jambhulkar et al. 2018). Similarly, bioenergy candidate plant species like *Saccharum munja*, *Jatropha curcas*, *Leucaena leucocephala* could also be grown in fly ash-amended soil. Apart from agroforestry, biomass and biofuel crops production, there is also opportunity to couple agricultural production on fly ash-amended soil with biofortification to improve the nutrient content of crops (Abhilash et al. 2016). Fly ash is also a good soil conditioner. Fly ash amendment has shown to increase the activity of enzymes that are key to the soil fertility. A 10% application of fly ash was reported to increase the activities of urease, dehydrogenase and phosphatase enzymes and improve the N cycle and P cycle in the soil. The increase in microbial activity was due to increased microbial biomass (Lai et al. 1999). However, application of fly ash at higher rate shows an adverse effect on soil microbial community and its associated functions. Fly ash also helps in buffering the soil pH. Fly ash of alkaline nature has been used for neutralizing the acidic soil by raising its pH (Saraswat and Chaudhary 2014). The lime component of the fly ash promotes the release of nutrients like B, S and Mo by reacting with acidic soil components. Fly ash application helps in improving the physical properties of soil like water holding capacity and the bulk density. Soil having poor soil texture with coarse particles can be amended with fly ash to improve the soil condition through reduced porosity, bulk density and increased water holding capacity and soil aggregate formation. Fly ash can be efficiently used in agricultural pest and disease control (Pandey et al. 2009; Mendki et al. 2001; Narayanasamy and Gnanakumar 1989; Sankari and Narayanasamy 2007; Eswaran and Manivannan 2007). Potential uses of fly ash in agriculture are depicted in Fig. 6.

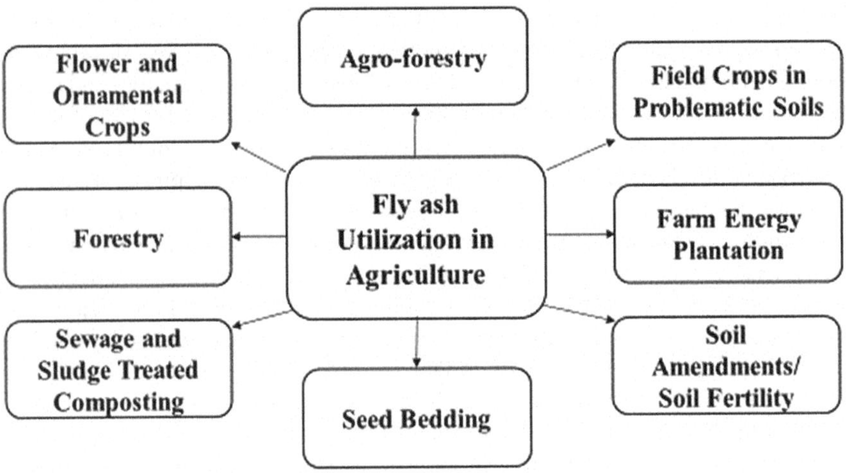

Fig. 6 Fly ash utilization in agriculture

3.3 *Seed Bedding*

Fly ash being a rich source of essential and micronutrients can be utilized as seed bedding material. Mupambwa et al. (2016) reported maximum number of flowers in marigold with the application of 25% fly ash and 75% pine bark as seed bedding material. In an experiment conducted by Jala and Goyal (2010), it was observed that mixing fly ash with soil on a w/w basis at 20% concentrations enhanced biomass production and plant growth in *Populus deltoides*. Application of fly ash at 10–20% level (v/v) improved biomass production by 20% in *Eucalyptus tereticornis* when used in nurseries as a potting mix material (Goyal et al. 2002). Banerjee et al. (2001) reported considerable improvement in germination, shoot height, leaf weight, root weight and total biomass of *Acacia auriculiformis* and *Azadirachta indica* in a media comprising soil + fly ash (1:1) and sand + soil + fly ash (1:1:1), respectively.

3.4 *Farm Energy Crops*

Bioenergy plant cultivation and energy production from various thermochemical processing of the phytobiomass can be regarded as a great tool for the utilization of fly ash. This also helps in production of carbon neutral fuel in comparison to its counterpart produced from fossil hence also minimizes the impact of climate change (Roy et al. 2018). Results of a study conducted by Mohan (2011) revealed that the addition of 20% fly ash in soil helped in increasing the chlorophyll content and photosynthetic rate in *Jatropha curcas* while Chaudhary and Ghosh (2013) reported significant improvement in biomass production and height of the tree with 5% fly ash addition to soil. Masto et al. (2013) observed improvement in maize yield, soil enzyme activity and microbial biomass with the conjunctive use of lignite fly ash and biochar. Xu et al. (2012) documented that the application of 6% fly ash and 14% sludge mixture registered maximum yield and nutrient concentration in Manila grass.

3.5 *Sewage and Sludge Composting*

Very common material used is lime for sewage sludge stabilization, which has a substantial part to play in lowering the microbial count of sludge-based pathogens as well as the heavy metal availability, and thus reduction of environmental risks and enhances the agricultural benefits (Samaras et al. 2008). Considering from environmental point of view, sustainable and safe application of sewage sludge/fly ash in conjunction on cultivated lands is recommended as valuable effort (Schoeman and Van Deventer 2004). Fly ash, instead of lime, composted with sewage sludge could augment the physical properties of soil adequately for use in agricultural and horticultural applications due to its enhanced presence as alternate source for providing nutrients to plants without affecting the soil conditions much though the presence of pathogenic organisms along with other pollutants, toxic metals, heavy metals and dissolvable salts present in the sewage sludge effect the soil conditions negatively. Towards the stabilization process, recent increased attention using suitable additives to prevent or suppress or minimize mobility or propagation of heavy metals in acquiescence to strict regulations by advanced countries is notable. Long-duration storage of lime sludge mixture resulting high pH value can establish drop of microbes and feeble metal bioavailability. Still, certain constraints like available time for stabilization and entailing cost associated are proposing the need of using other materials to take the role. Fly ash probably rich in oxides of calcium and magnesium relying on the source of coal used in the power plants from thermal power plant may have similarity with lime, reduces heavy metal availability by physical adsorption and precipitation at high pH.

3.6 Forestry and Agroforestry

A number of forest species like *Acacia* sp., *Albizia lebbek*, *Dalbergia sissoo*, *Eucalyptus hybrid* and *Tamarindus indica* have been grown on soils amended with fly ash for restoration of fly ash dump sites. Trees like *Dalbergia sissoo*, *Populus deltoides*, *Melia azedarach*, *Eucalyptus* sp. (Adholeya et al. 1998), *Grevillea pteridifolia*, *Gmelina arborea* and *Emblica officinalis* (Banerjee et al. 2003) established well on fly ash overburdens. An enhancement of 10% in growth of, *Acacia auriculiformis*, *Casuarina equisetifolia and Eucalyptus tereticornis* was observed with utilization of ESP fly ash at 18–24% (v/v) (Goyal et al. 2002). In another study conducted in Florida by Riekerk (1984) on poorly drained sandy soil, fly ash addition altered the run off pH from 4.7 to 6.0 and contributed to enhanced growth rate in Australian pine plantation. 66% amendment of fly ash by soil volume of the planting pit considerably augmented the teak diameter at breast height and also reduced the bulk density and increased plant available water in teak (Ramesh et al. 2008). Addition of fly ash in conjunction with pressmud increased plant growth plus other physiological parameters like protein, chlorophyll and in vivo activity of nitrate reductase in subabul (Gupta et al. 2000).

3.7 Flower and Ornamental Crops

Fly ash can be utilized in floriculture/ornamental plant production as it is a source of essential nutrients required for plant growth and development. The presence of toxic heavy metals in fly ash restricts its use in edible crops. Hence, utilization of fly ash in floriculture/ornamental plant cultivation can be regarded as a better option compared to its use in edible crops. Comparable fresh and dry weights, canopy widths or quality grading were observed in plants grown on substrates treated with fly ash or dolomite revealing that fly ash could be a substitute dolomite for containerized lily production (Chen and Li 2006). Menzies and Aitken (1996) evaluated the growth parameters of Petunia in fly ash-treated substrates and suggested that fly ash can be utilized as container substrates component effectively for ornamental plant production. In another study carried out by Tripathi et al. (2007), the addition of distillery discharge with fly ash in the ratio of 1:2 to soil as a manure enhanced the growth and productivity of Gladiolus. Ray et al. (2012) reported that the vegetative growth parameters, viz. root length, leaves and leaf area, number of branches, shoot length and number of flowers in Calendula have increased in soils treated with fly ash indicating that a mixing fly ash and distillery effluent can significantly improve the vegetative growth and synthesis of chlorophyll.

4 Fly Ash for Land Restoration and Environmental Services

4.1 Reduced Lime Utilization

In India, there are 90 m ha of acidic soils, of which 31 m ha are severely acidic with $pH < 5.5$. In severely acidic soils, application of lime to raise pH to neutral level (>6.0 pH) and also to avoid manganese and aluminium toxicity problems is desired. If we assume 2.0 t/ha of lime is used per annum on 31 m ha strongly acidic soil, requirement of lime calculated to be 62 mt lime/year. Moreover, fly ash is also a potential liming material which neutralizes the acidity of the soil (Taylor and Schumann 1988), in stabilization of sewage sludge when utilized as a source of eco-remediation at mining sites (Zhang et al. 2008). Addition of fly ash to concrete for construction and road works also reduces the use of lime.

4.2 CO_2 Emission Reduction

In the global emissions of the GHGs like CO_2, N_2O and CH_4, agriculture plays a conspicuous role. Studies conducted by various researchers suggested that new opportunities have appeared for minimizing the global warming potential by modifying the agronomic practices (Robertson et al. 2000). The Intergovernmental Panel on Climate Change (IPCC) assumed that entire carbon of agricultural lime is ultimately released into the atmosphere as CO_2. Further USEPA showed that 9 Teragram of CO_2 was released from 20 Tg of used agricultural lime in 2001 (West and McBride 2005). Montes-Hernandez et al. (2009) from experimental studies have arrived that 38.18 t fly ash use can mitigate a ton of CO_2 emission. Fly ash utilization as a soil ameliorant substituting lime might lead to CO_2 emission reduction, thus contribute to minimize global warming. Similarly, it was estimated that use of a ton of fly ash amounts to 2 tons of CO_2 emission reduction from way of reduced cement production (Krishnamoorthy 2000; Naik and Tyson 2000).

4.3 Restoration of Degraded Lands

Extreme variability in pH of fly ash varying from highly acidic to alkaline offers for it wide utilization in amending alkaline/saline/waterlogged/acid soils (Malik and Thapliyal 2009). Fly ash has been shown to neutralize soil acidity by acting as a liming material (Taylor and Schumann 1988). The soil acidity neutralizing capacity of fly ash emanates from the hydroxide and carbonate salts (Matsi and Keramidas 1999; Pathan et al. 2003; Cetin and Pehlivan 2007). Owing to alkaline nature of major fraction of fly ash produced in India, its soil application could elevate the soil pH and thus neutralize acidity (Phung et al. 1978). $CaCO_3$ (lime) equivalent of fly ash will differ depending on its composition. McCarty et al. (1994) have given a $CaCO_3$ equivalence of 18 and 97 for fly ash and bed ash for USA. Application of 1120 Mg/ha has enhanced the 0–15 cm soil pH from 4.89 to 6.45 within a year after application (Adriano et al. 2002). The pH of fly ash was reported to vary from 4.5 to 12.0 based on the sulphur level in the coal (Adriano et al. 1980). Thus, fly ash with high pH could be used in acidic soils as a liming material. Not only the moderation of soil pH to near neutral level, but also the increased nutrient availability of calcium and magnesium and overcoming of toxic effects of aluminium and manganese (Fail and Wochok 1977; Shende et al. 1994) with application of fly ash makes it more than a liming material and thus contributes to greater enhancements in productivity of acidic soils.

Gypsum was suggested to be substituted by 40% of its requirement with moderately very less acidic fly ash for reclaiming sodic soils (Kumar and Singh 2003). In sodic soils, fly ash proved beneficial in improving physico-chemical properties which ultimately resulted in considerable increase in wheat and rice production. Addition of fly ash and pyrite amendment has lowered the pH of the soil from 10.0 to 8.3. Also, the sodium saturation was altered from 64.5 to 24.0 (Tiwari et al. 1992). Therefore, soil pH amelioration utilization fly ash has a major role in transforming huge areas including waste lands, non-productive, mine spoils, etc. under farming and forestry purposes.

Since fly ash also found to have various nutrients essential for the plant growth (for example, Mg, K, Ca, B, Fe, S, Cu, P and Zn) (Sajwan et al. 2003), it can be utilized as an aid for nutrient supplementation for the plants mediated revitalization of the polluted, marginal and degraded lands. Purposeful and wise land application of fly ash offers additional avenues and promising opportunities, particularly related towards the pH optimization (Ciccu et al. 2003; Shaheen and Tsadilas 2010) and improving soil physical conditions, viz. water holding capacity, water infiltration, bulk density, hydraulic conductivity and soil aggregation (Pathan et al. 2003; Singh et al. 2010;

Ukwattage et al. 2013; Skousen et al. 2013). Improvement in soil compaction and aggregate stability augmentation in sodic soils can be attained by the application of fly ash (Kumar and Singh 2003). Substantial sorption potential for organic pollutants has been exhibited by fly ash (Konstantinou and Albanis 2000); thus, its soil application might assist in reducing the plunging movement of soil-applied pesticides, particularly the herbicides (Ghosh and Singh 2012). One of the herbicides used for pre- and post-emergence regulator of broadleaf weeds and annual grasses in the cultivation of soybean, wheat, sugarcane and other cash crops is metribuzin. It is feebly sorbed in the soil and thus shows lateral as well as downward mobility in soil column (Singh and Raunaq Singh 2013a, b). Majumdar and Singh (2007) studied the impact of fly ash amendments at two levels [2.5% (fly ash 1) and 5% (fly ash 2), w/w] on the mobility and sorption of metribuzin in the soil system. Mass balance equation showed that only 26% of the initially applied metribuzin was recovered in the leachate in fly ash-amended soil column. Moreover, almost no metribuzin was recovered in the leachate in another soil column. Further, it was shown that fly ash amendment helped in reducing the metribuzin loss due to leaching by 75–100% which clearly shows that fly ash application can enhance the metribuzin retention in the soil (Majumdar and Singh 2007). Furthermore, research studies also depict that incorporating alkaline fly ash with or without other amendments can play a vital role in restoring marginally metal-contaminated soils via immobilization of mobile metal moieties. Fly ash was applied to restore a Cu- and Pb-polluted soil (Kumpiene et al. 2007). The study indicated that fly ash-amended soil lowered the discharge of Cu and Pb by around 96–99.9% after two years of time period. Such amendment helps in lowering the exchangeable metal moieties due to the establishment of new inorganic Cu and Pb capturing phases and an escalated sorption of metal because of higher number of sorption sites (Kumpiene et al. 2007).

4.4 *Saving of Productive Land from Brick Kiln Making by Shifting to Fly Ash Bricks*

Globally, India ranks second in brick kiln production (approx. 140 billion bricks per year) after China. Brick manufacturing is an unorganized, conventional industry restricted to peri-urban and rural areas. The Gangetic plain contributes to about 65% of the total brick production. The major brick producing states in this region are Punjab, Haryana, Uttar Pradesh, Bihar and West Bengal (Singh and Asgher 2005). Resorting to fly ash brick making in lieu of earthen bricks in kilns has twin benefits. It solves the disposal problems of fly ash and the land required for their disposal (2072 ha/annum). Also, the precious soil utilized for making bricks can be saved by resorting to fly ash bricks.

4.5 Other Uses of Fly Ash

Apart from the uses discussed above, fly ash can be used in production of silicon-based fertilizers as silicon is a major component of fly ash, as filler material for fertilizer, granular pesticides and dust formulation type of pesticides, as an absorbent of moisture in water logged soils, barnyard paving device, in synthesizing fly ash-based composite polymer for wood substitute and manufacturing of alum, domestic cleaning powder.

5 Limitations of Fly Ash Application

Difficulties in fly ash utilization for soil amendment arise with the adverse alteration in the soil pH, disordered supply of nutrients, boron toxicity for plants, excessive supply of sulphate as well as other toxic elements. Such adverse effects are generally observed under inappropriate or excess applications of fly ash. The levels of POPs such as PAHs and PCBs in fly ash are usually low; furthermore, monitoring studies should be prioritized regarding their impacts on soil biota, nutrients uptake by plants and soil retention. Though, fly ash application with sewage sludge or manure to land enhances the quality of the soil, it poses threat to the safety of soil ecosystem due to presence of pathogenic microorganisms and other contaminants in the sewage sludge. Other major limitations of application are uptake and accumulation of heavy metals and radionuclide in plants growing in fly ash-contaminated/amended soil. It may lead to contamination of the food chain upon consumption causing biomagnification mediated deleterious effect on human beings and animals. Higher concentration of As, B, Se, Mo and Sr was reported in plant tissues of plants grown in fly ash-treated soil (Adriano et al. 2002). The presence of highly toxic heavy metals (As, Hg, Pb, Zn, Ni, V, Cr, Cd, etc.) and organic pollutants (Polyaromatic hydrocarbon (PAH), Polychlorinated biphenyl (PCB), Polychlorinated dibenzo-p-dioxins (PCDD), Polychlorinated dibenzofurans (PCDF), etc.) in fly ash (Sahu et al. 2004) poses threat to soil, groundwater and vegetation contamination. Further phytotoxicity of these contaminants limits the growth of plant in fly ash-contaminated/treated soil. Fly ash is often rich in boron, and boron toxicity significantly reduces the crop yield. The leaching and mobility of heavy metals from fly ash-amended soil may lead to contamination of pristine ecosystems through source point contamination. Also, there is a deficiency of Zn and P in fly ash. Further, application of fly ash is known to increase the soil and groundwater salinity. Moreover, the biotechnological advances like new omics revolution, nanotechnological developments and the discovery of novel catabolic pathways can help in enhancing the efficacy of fly ash remediation and augment its applicability to clean up such contaminated soils (Tripathi et al. 2017).

6 Conclusion

There is a general consensus that fly ash is an environmental hazard. A large amount of fly ash produced in the coal-based thermal energy production process creates a serious problem for its safe disposal. Despite the toxicity, the multi-nutrient nature of fly ash provides ample opportunity for its utilization in the agriculture sector. Research has proved that fly ash can be applied in safer limits for increasing crop productivity, improving soil physico-chemical and biological properties and soil nutrient replenishment. Though the heavy metals, radionuclide and organic pollutant content of the fly ash limit its utilization for edible crop production due to food chain contamination, however, it could be utilized in agroforestry and biomass and bioenergy crop production. Further, nutrient-rich fly ash can be used for supporting plant growth in the degraded and contaminated land for its successful restoration.

7 Way Forward

- Integrated use of fly ash along with organic sources or fertilizers would help in lowering the heavy metal toxicity and pave way for its increased utilization in food crops
- Fly ash dump sites could be planted with tolerant bioenergy plants for restoration coupled bioenergy production
- Biomonitoring of the crops growing in fly ash-amended/contaminated soil is necessary to restrict food chain contamination
- More fly ash-tolerant indigenous plants should be explored and beneficial plant–microbe interactions should be exploited for robust plant production in fly ash-contaminated soil
- The primer driver of fly ash utilization should always be regulatory compliance rather profit generation.

References

Abhilash, P. C., Tripathi, V., Edrisi, S. A., Dubey, R. K., Bakshi, M., Dubey, P. K., et al. (2016). Sustainability of crop production from polluted lands. *Energy, Ecology, and the Environment, 1*(1), 54–65.

Adholeya, A., Bhatia, N. P., Kanwar, S., & Kumar, S. (1998). Fly ash source and substrate for growth of sustainable agro-forestry system. In *Proceedings of Regional Workshop cum Symposium on Fly Ash Disposal and Utilization*. Organized by Kota Thermal Power Station, RSEB, Kota, and Rajasthan, India.

Adriano, D. C., Weber, J., Bola, N. S., Paramasivam, S., Koo, B-J, & Sajwan, K. S. (2002). Effects of high rates of coal fly ash on soil, turfgrass, and groundwater quality. *Water, Air, and Soil pollution, 139*, 365–385.

Adriano, D. C., Page, A. L., Elseewi, A., Chang, A. C., & Straughan. (1980). Utilization and disposal of fly ash other coal residues in terrestrial ecosystems: A review. *Journal of Environmental Quality, 9,* 333–344.

Ahmaruzzaman, M. (2010). A review on the utilization of flyash. *Progress in Energy and Combustion Science, 36,* 327–363.

Asokan, P., Saxena, M., & Bose, S. K. Z. (1995). In *Proceedings of the workshop on Flyash Management in the State of Orissa* (pp. 64–75). Bhubaneshwar, India: RRL.

Banerjee, S., Singh, A. K., & Banerjee, S. K. (2003). Impact of fly ash on foliar chemical and biochemical composition of naturally occurring ground flora and its possible utilization for growing tree crops. *Indian Forester, 129,* 964–978.

Banerjee, S. K., Singh, A. K., Jain, A., Bhowmik, A. K., & Manjhi, R. B. (2001). *Pollution absorbing efficiency of tree species in industrial area* (85–89). Final report: Fly ash Project. Jabalpur: TFRI Publication.

Basu, M., Bhadoria, P. B. S., & Mahapatra, S. C. (2007). Role of soil amendments in improving groundnut productivity of acid lateritic soils. *International Journal of Agricultural Research, 2*(1), 87–91.

CEA. (2016–17). Report on fly ash generation at *Coal/Lignite Based Thermal Power Stations and its Utilization in the Country for the year 2016–17.* Central Electricity Authority, New Delhi, 2017.

Cetin, S., & Pehlivan, E. (2007). The use of fly ash as a low cost, environmentally friendly alternative to activated carbon for the removal of heavy metals from aqueous solutions. *Colloids and Surfaces A, 298,* 83–87.

Chandra, A., & Chandra, H. (2004). Impact of Indian and imported coal on Indian thermal power plants. *Journal of Scientific & Industrial Research, 63,* 156–162.

Chang, A. C., Lund, L. J., & Page, A. L. (1977). Physical properties of flyash amended soils. *Journal of Environmental Quality, 6,* 267–270.

Chaudhary, D. R., & Ghosh, A. (2013). Bioaccumulation of nutrient elements from fly ash-amended soil in *Jatropha curcas* L, a biofuel crop. *Environmental Monitoring and Assessment, 185,* 6705–6712.

Chen, J., & Li, Y. (2006). Coal fly ash as an amendment to container substrate for *Spathiphyllum* production. *Bioresource Technology, 97,* 1920–1926.

Ciccu, R., Ghiani, M., Serci, A., Fadda, S., Peretti, R., & Zucca, A. (2003). Heavy metal immobilization in the mining-contaminated soils using various industrial wastes. *Minerals Engineering, 16,* 187–192.

Costa, B. M. (2016). *Sintese de zeolitasa partir de cinzas de carvao para captura de dioxide de carbono.* UFRS, Porto Alegre, RS: Mestrado dissertacao.

Dwivedi, S. F., Tripathi, R. D., Srivastava, S., Mishra, S., Shukla, M. K., Tiwari, K. K., et al. (2007). Growth performance and biochemical responses of three rice (*Oryza sativa* L.) cultivars grown in fly-ash amended soil. *Chemosphere, 67*(1), 140–151.

Eswaran, A., & Manivannan, K. (2007). Effect of foliar application of lignite fly ash on the management pf papaya leaf curl disease. *Acta Horticulturae (ISHS), 740,* 271–275.

Fail, J. L., Jr., & Wochok, Z. S. (1977). Soybean growth on fly ash amended strip mine spoils. *Plant and Soil, 48,* 473–484.

Feng, Y. J., Li, F., Wang, X. L., Liu, X. M., & Zhang, L. N. (2006). Principal chemical properties of artificial soil composed of fly ash and furfural residue. *Pedosphere, 16*(5), 668–672.

Furr, A. K., Parkinson, T. F., & Gutenmann, W. H. (1978). Elemental content of vegetables, grains and forages field grown on flyash amended soils. *Journal of Agriculture and Food Chemistry, 26,* 357–359.

Garg, R. N., Pathak, H., & Das, D. K. (2005). Use of flyash and biogas slurry for improving wheat yield and physical properties of soil. *Environmental Monitoring and Assessment, 107,* 1–9.

Ghosh, R. K., & Singh, N. (2012). Managing metolachlor and atrazine leaching losses using lignite fly ash. *Ecotoxicology and Environmental Safety, 84,* 243–248.

Goyal, D., Kaur, K., Garg, R., Vijayan, V., Nanda, S. K., Nioding, A., et al. (2002). Industrial fly ash as a soil amendment agent for raising forestry plantations. In P. R. Taylor (Ed.), *EPD congress and fundamental of advanced materials for energy conversion* (pp. 251–260). Warrendale, PA: TMS Publication.

Gupta, M., Kumar, K., & Yunus, M. (2000). Effect of fly ash on metal composition and physiological responses in *Leucaena leucocephala* (LAMK.) DE.WIT. *Environmental Monitoring and Assessment, 61,* 399–406.

Gupta, A. K., & Sinha, S. (2006). Role of *Brassica juncea* (L.) Czern. (var. Vaibhav) in the phytoextraction of Ni from soil amended with fly ash: Selection of extractant for metal bioavailability. *Journal of Hazardous Materials, 136*(2), 371–378.

Jala, S., & Goyal, D. (2010). Fly ash application effects on plant biomass and bio-concentration of micronutrients in nursery seedlings of *Populus deltoides*. In 19th World congress of soil science: Soil solutions for a changing world. *Procedings Wein, International Union of Soil Sciences Symposium* (pp. 53–56). Brisbane, 1–6 August 2010.

Jambhulkar, H. P., Shaikh, S. M., & Kumar, M. S. (2018). Fly ash toxicity, emerging issues and possible implications for its exploitation in agriculture; Indian scenario: A review. *Chemosphere, 213,* 333–344.

Konstantinou, I. K., & Albanis, T. A. (2000). Adsorption desorption studies of selected herbicides in soil-fly ash mixtures. *Journal of Agriculture and Food Chemistry, 48,* 4780–4790.

Krishnamoorthy, R. (2000). Ash utilization in India—Prospect and problems. Barrier and utilization option of large volume application of fly ash in India. In V. Hajela (Ed.), *Proceedings of the Workshop on USAID/India Greenhouse Gas Pollution Prevention Project* (pp. 63–67).

Kumar, D., & Singh, B. (2003). The use of coal fly ash in sodic soil reclamation. *Land Degradation and Development, 14,* 285–299.

Kumpiene, J., Lagerkvist, A., & Maurice, C. (2007). Stabilization of Pb and Cu contaminated soil using coal fly ash and peat. *Environmental Pollution, 145,* 365–373.

Lai, K. M., Ye, D. Y., & Wong, J. W. C. (1999). Enzyme activities in a sandy soil amended with sewage sludge and coal fly ash. *Water, Air, and Soil Pollution, 113,* 261–272.

Lau, S. S. S., & Wong, J. W. C. (2001). Toxicity evaluation of weathered coal fly ash amended manure compost. *Water, Air, and Soil Pollution, 128,* 243–254.

Majumdar, K., & Singh, N. (2007). Effect of soil amendments on sorption and mobility of metribuzin in soils. *Chemosphere, 66,* 630–637.

Malik, A., & Thapliyal, A. (2009). Eco-friendly fly ash utilization: Potential for land application. *Critical Reviews in Environmental Science and Technology, 39,* 333–366.

Masto, R. E., Mohammad, A., George, J., Selvi, V. A., & Ram, L. C. (2013). Co-application of biochar and lignite fly ash an soil nutrients and biological parameters at different crop growth stages of Zea mays. *Ecological Engineering, 58,* 314–322.

Matsi, T., & Keramidas, V. Z. (1999). Flyash application on two acid soils and its effect on soil salinity, pH, B, P and on ryegrass growth and composition. *Environmental Pollution, 104,* 107–112.

Mattigod, S. V., Dhanpat, R., & Eary, L. E. (1990). Geochemical factors controlling the mobilization of inorganic constituents from fossil fuel combustion residues: I. Review of the major elements. *Journal of Environmental Quality, 19,* 188–201.

McCarty, G. W. R., Siddaramappa, R. J., Wright, E. E., & Gao, G. (1994). Evaluation of coal combustion byproducts as soil liming materials: Their influence on soil pH and enzyme activities. *Biology and Fertility of Soils, 17,* 147–172.

Mendki, P. S., Maheshwari, V. L., & Kothari, R. M. (2001). Fly-ash as a post-harvest preservation for five commonly utilized pulses. *Crop Protection, 20,* 241–245.

Menzies, N. W., & Aitken, R. L. (1996). Evaluation of fly ash as component of potting substrates. *Scientia Horticulturae, 67,* 87–99.

Mohan, S. (2011). Growth of biodiesel plant in fly ash: A sustainable approach response of *Jatropha curcas*, a biodiesel plant in fly ash amended soil with respect to pigment content and photosynthetic rate. *Procedia Environmental Sciences, 8,* 421–425.

Montes-Hernandez, G., Perez-Lopez, R., Renard, F., Nieto, J. M., & Charlet, I. (2009). Mineral sequestration of CO_2 by aqueous carbonation of coal combustion fly ash. *Journal of Hazardous Materials, 16,* 1347–1354.

Mupambwa, H. A., Lukashe, N. S., Nyari, P., & Mnkeni, S. (2016). Suitability of fly ash vermicompost as a component of pine bark growing media: Effects on media physicochemical properties and ornamental marigold growth and flowering. *Compost Science & Utilization.*

Naik, T. R., & Tyson, S. S. (2000). Environmental benefits from the use of coal combustion products (CCP). In C. V. J. Verma, S. V. Rao, V. Kumar, & R. Krishnamoorthy (Eds.), *Proceedings of the Second International Conference on Fly ash Disposal and Utilization* (pp. 40–43).

Narayanasamy, P., & Gnanakumar, D. (1989). A lignite fly-ash: A nonpolluting substance for tackling pest problems. In K. V. Devaraj (Ed.), *Progress in pollution research* (pp. 201–206). University of Agricultural Science, Bangalore.

Natusch, D. F. S., & Wallace, J. R. (1974). Urban aerosol toxicity: The influence of particle size. *Science, 186,* 695.

Neelima, M. R., Khandual, S., & Tripathy, A. (1995). In *Proceedings of Workshop on Fly ash Management in the State of Orissa, April 11* (pp. 76–89). Bhubhaneshwar, India: RRL.

Pandey, V. C., Abhilash, P. C., & Singh, N. (2009). The Indian perspective of utilizing fly ash in phytoremediation, phytomanagement and biomass production. *Journal of Environmental Management, 90,* 2943–2958.

Pandey, V. C., & Singh, N. (2010). Impact of fly ash incorporation in soil systems. *Agriculture, Eco-systems and Environment, 136,* 16–27.

Pathan, S. M., Aylmore, L. A., & Colmer, T. D. (2003). Soil properties and turf growth on a sandy soil amended with fly ash. *Plant and Soil, 256,* 103–114.

Phung, H. T., Lund, I. J., & Page, A. L. (1978). Potential use of fly ash as a liming material. In D. C. Adriano & I. L. Brisbin (Eds.), *Environmental chemistry and cycling processes, Conf-760429* (pp. 504–515). Springfield, VA: US Department of Commerce.

Ramesh, V., Korwar, G. R., Mandal, U. K., Prasad, J. V. N. S., Sharma, K., Ramakrishna, S., & Kandula, V. (2008). Influence of fly ash mixtures on early tree growth and physicochemical properties of soil in semi-arid tropical alfisols. *Agroforestry Systems, 73,* 13–22.

Ray, S. N. C., Ranjan, N., Kumari, K., & Sinha, R. C. (2012). Potential use of mix of fly ash and diluted distillery effluent in agriculture: A case study on the vegetative plant, *Calendula officinalis. International Journal of Engineering Research and Applications (IJERA), 2248–9622,* 193–203.

Riekerk, H. (1984). Coal–ash effects on fuelwood production and runoff water quality. *Southern Journal of Applied Forestry, 8,* 99–102.

Robertson, G. P., Paul, E. A., & Harwood, R. R. (2000). Greenhouse gases in intensive agriculture: Contributions of individual gases to the radiative forcing of the atmosphere. *Science, 289*(5486), 1922–1925.

Roy, M., Roy Chowdhury, R., & Mukherjee, P. (2018). A review on remediation of fly ash dumpsites through bioenergy crop plantation and generation. *Pedosphere, 28*(4), 561–580.

Roy, W. R., Thiery, R. G., & Schuller, R. M. (1981). Coal flyash: A review of the literature and proposed classification system with emphasis on environmental impacts. *Environmental Geology Notes 96*. Champaign, IL: Illinois State Geological Survey.

Sahu, S. K., Pandit, G. G., & Sadasivan, S. (2004). Precipitation scavenging of polycyclic aromatic hydrocarbons in Mumbai, India. *Science of the Total Environment, 318*(1–3), 245–249.

Sajwan, K. S., Paramasivam, S., Alva, A. K., Adriano, D. C., & Hooda, P. S. (2003). Assessing the feasibility of land application of fly ash, sewage sludge and their mixtures. *Advances in Environmental Research, 8,* 77–91.

Samaras, P., Papadimitriou, C. A., Haritou, I., & Zouboulis, A. I. (2008). Investigation of sewage sludge stabilization potential by the addition of fly ash and lime. *Journal of Hazardous Materials, 154,* 1052–1059.

Sankari, S. A., & Narayanasamy, P. (2007). Bio-efficancy of fly-ash based herbal pesticides against pests of rice and vegetables. *Current Science, 92,* 811–816.

Sao, S., Gothalwal, R., & Thetwarj, L. K. (2007). Effects of flyash and plant hormones treated soil on the increased protein and amino acid contents in the seeds of ground nut (*Arachis hypogaea*). *Asian Journal of Chemistry, 19*(2), 1023–1026.

Saraswat, P. K., & Chaudhary, K. (2014). Effect of Fly Ash (FA) improving soil quality and increase the efficiency of crop productivity. *European Journal of Biotechnology and Bioscience, 2*(6), 72–78.

Schoeman, J. L., & van Deventer, P. W. (2004). Soils and the environment: The past 25 years. *South African Journal of Plant and Soil, 21*(5), 369–387.

Sett, R. (2017). Flyash: Characteristics, problems and possible utilization. *Advances in Applied Science Research, 8*(3), 32–50.

Shaheen, S. M., & Tsadilas, C. D. (2010). Sorption of cadmium and lead by acidic Alfisols as influenced by fly ash and sewage sludge application. *Pedosphere, 20,* 436–445.

Shende, A., Juwarkar, A. S., & Dara, S. S. (1994). Use of fly ash in reducing heavy metal toxicity in plants. *Resources, Conservation and Recycling, 12,* 221–228.

Singh, A. L., & Asgher, M. S. (2005). Impact of brick kilns on land use/landcover changes around Aligarh city, India. *Habitat International, 29*(3), 91–602.

Singh, R. P., Gupta, A. K., Ibrahim, M. A., & Mittad, A. K. (2010). Coal fly ash utilization in agriculture: Its potential benefits and risks. *Reviews in Environmental Science & Biotechnology, 9,* 345–358.

Singh, N., & Raunaq Singh, S. B. (2013a). Reduced downward mobility of metribuzin in fly ash-amended soils. *Journal of Environmental Science and Health. Part B: Pesticides, Food Contaminants, and Agricultural Wastes, 48,* 587–592.

Singh, N., & Raunaq Singh, S. B. (2013b). Effect of fly ash amendment on persistence of metribuzin in soils. *Journal of Environmental Science and Health. Part B: Pesticides, Food Contaminants, and Agricultural Wastes, 48,* 108–113.

Skousen, J., Yang, J. E., Lee, J., & Ziemkiewicz, P. (2013). Review of fly ash as a soil amendment. *Geosystem Engineering, 16,* 249–256.

Szponder, D. K., & Trybalski, K. (2009). Identification of the fly ash properties by using different methods and equipments (Okreslanie wlasciwosci popiolow lotnych przy uzyciu roznych metod i urzadzen badawczych). *Geoinzynieria, 33*(4), 287–298.

Taylor, E. M., & Schumann, G. E. (1988). Flyash and lime amendment of acidic coal soil to aid revegetation. *Journal of Environmental Quality, 17,* 120–124.

Thetwar, L. K., Sahu, D. P., & Vaishnava, M. M. (2006). Analysis of *Sesamum indicum* seed oil from plants grown in flyash amended acidic soil. *Asian Journal of Chemistry, 18*(1), 481–484.

Tiwari, K. N., Sharma, D. N., Sharma, V. K., & Dingar, S. M. (1992). Evaluation of fly ash and pyrite for sodic soil rehabilitation in Uttar Pradesh, India. *Arid Soil Research and Rehabilitation, 6,* 117–126.

Tolle, D. A., Arthur, M. F., & Pomeroy, S. E. (1982). Flyash use for agriculture and land reclamation: A critical literature review and identification of additional research needs. RP-1224-5. Columbus, Ohio: Battelle Columbus Laboratories; 1982. *Environmental Monitoring and Assessment, 107,* 1–9.

Tripathi, V., Edrisi, S. A., Chen, B., Gupta, V. K., Vilu, R., Gathergood, N., et al. (2017). Biotechnological advances for restoring degraded land for sustainable development. *Trends in Biotechnology, 35*(9), 847–859.

Tripathi, S., Joshi, H. C., Sharma, D. K., & Singh, K. P. (2007). Effect of distillery effluent and flyash mixture on soil fertility, plant growth and flower yield in Gladiolus. *Journal of Ornamental Horticulture, 2007–10*(1), 34–37.

Ukwattage, N. L., Ranjith, P. G., & Bouazza, M. (2013). The use of coal combustion fly ash as a soil amendment in agricultural lands (with comments on its potential to improve food security and sequester carbon). *Fuel, 109,* 400–408.

West, T. O., & McBride, A. C. (2005). The contribution of agricultural lime to carbon dioxide emissions in the United States: Dissolution, transport, and net emissions. *Agriculture, Ecosystems & Environment, 108*(2), 145–154.

Xu, J. Q., Yu, R.-L., Dong, X.-Y., Hu, G.-R., Shang, X.-S., Wang, Q., et al. (2012). Effects of municipal sewage sludge stabilized by fly ash on the growth of Manilagrass and transfer of heavy metals. *Journal of Hazardous Materials, 217–218,* 58–66.

Zhang, H., Sun, L., & Sun, T. (2008). Solubility of ion and trace metals from stabilized sewage sludge by fly ash and alkaline mine tailing. *Journal of Environmental Sciences, 20,* 710–716.

Carbon and Nutrient Sequestration Potential of Coal-Based Fly Ash Zeolites

V. Ramesh and James George

Abstract Disposal, safe management, and gainful utilization of coal-based fly ash are the issues of major concern and challenge in the present century due to alarming increase in the production of ash in India Recent reports indicated that fly ash utilization in agriculture sector has stood at 1.92 mt (million tones) during 2016–17, which constitutes hardly 1.14% alone of the total fly ash utilization. This might be attributed to low product value and presence of heavy metals in fly ash which limits its large-scale applications as agricultural soil amendments. Conversion of fly ash into zeolites (FAZ), a sodium aluminosilicates group of minerals, is an innovative and proven approach but not adequately researched under laboratory conditions to engineer the right quality of zeolite and field conditions to find out the slow nutrient release characteristics and the use efficiency under diverse soil types and agroecological conditions. The improved percent zeolitization of the FAZ will have twin benefits (water and nutrient retention) because of its fine loamy texture and high cation-exchange capacity (CEC). In addition, zeolites can conserve soil organic matter that will help further to improve the efficiency of soil water and nutrient use. This added property will be highly beneficial for tuber crops in particular as they are extensively grown in the country in degraded and marginally fertile soils poor in soil organic carbon (SOC) and as the economic parts are beneath the soil, the physical properties and SOC content are critical for the crop performance. With the funding support of fly ash unit, DST, during 2010–13, low-cost, high-value agricultural grade fly ash zeolites (near-neutral pH, low Na, high CEC, and low heavy metals content) was successfully synthesized and evaluated in sweet potato (*Ipomea batatus* L.) wherein the tuber yield was found correlated in soils amended with zeolites due to higher nutrient use efficiency with respect to the major nutrients studied, viz. N, P, and K. Research on large-scale field application especially on soil aggregation and compaction properties, soil carbon quality and stability potential aspects of FAZ, adsorption and availability of important soil nutrients NH_4^+, K^+, Ca^{++}, and

---

V. Ramesh (✉)
ICAR-Central Tuber Crops Research Institute (CTCRI), Thiruvananthapuram, Kerala, India
e-mail: varadharajan_ramesh@rediffmail.com

J. George
AICRPTC, ICAR-CTCRI, Thiruvananthapuram, Kerala, India
e-mail: jgkarott@gmail.com

S. K. Ghosh and V. Kumar (eds.), *Circular Economy and Fly Ash Management*,
https://doi.org/10.1007/978-981-15-0014-5_4

Na^+ must be given utmost priority to further establish the controlled release fertilizer characteristics of FAZ in India.

Keywords Fly ash zeolites · Cation exchange · Soil amendment · Soil carbon · Sweet potato · Soil quality

1 Introduction

Potential applications of zeolites exist in agriculture in improving soil water and nutrient use efficiency of low-quality soils. In India, tuber crops are widely grown in low-quality soils with poor SOC content. This considerably decreased the soil physical capacity that affects water retention and nutrient supply to plants. This situation also necessitated farmers to apply high quantities of fertilizers for yield benefits that resulted in severe nutrient imbalance besides polluting soil, water, and atmosphere. Additions of more quantities of organic manures in view of diminishing cattle population are not a feasible option. Rather, conservation and efficient use of added organic matter is the only strategic option available in the present scenario. Hence, development of product and adoption of technologies to increase the resource use efficiency with minimum additions (concentrated form) without polluting the soil is the need of the hour. Addition of zeolites to soil helps to improve water holding and nutrient adsorption and ion exchange properties that will help in supplying plant nutrients more effective for plant growth. The advantages of natural zeolites such as clinoptilolite for soil applications are well established. Natural zeolites deposit are not available in India and imported zeolites are costlier. At present, synthetic Zeolites is widely used in detergents, water purification and as catalysts in petrochemical industry but the utility in agriculture is not intensively explored in India. The main advantages of synthetic zeolites as compared to natural zeolites are that they can be engineered with a wide variety of chemical properties and pore sizes and that they have greater thermal stability. Moreover, fly ash was an excellent raw material as compared to other sources, due to mineralogical resemblance of natural zeolites. Similar to natural zeolites, fly ash-based zeolites (FAZ) also possess the capacity to sequester ions in the lattice positions or within the network of their channels and voids.

But, FAZ for agricultural soil applications will be beneficial and have fertilizer value only if it possesses the following three characteristics, viz. low sodium content with near-neutral pH, high cation-exchange capacity (CEC), and maximum surface area for effective adsorption and retention of major cations of plant nutrient value and minimum heavy metal content. FAZ synthesis is conventionally developed by hydrothermal crystallization under alkaline conditions, which has been reported by several patents and technical articles. However, additions of higher quantities of FAZ may invite changes in soil physical atmosphere, which are very critical in tuber crops as the economic products are present beneath the soil. However, there is a limited understanding of the processes and mechanisms of producing zeolites from fly ash with the above three characteristics that will make this product more suitable and potential in agricultural soil applications. The effect of such zeolites additions on soil physicochemical properties, viz. soil aggregation, soil compaction, soil moisture, available soil nutrients, and plant growth parameters needs detailed work under field conditions. At ICAR-CTCRI, high cation-exchange capacity, low sodium-containing fly ash-based zeolites (zeolites mixtures, which contained more than one species of zeolites) suitable for agricultural soil applications was synthesized and evaluated in sweet potato under controlled pot culture conditions. Soils amended with 1% zeolites in pots considerably improved the nutrient uptake of NPK and tuber yield (57%) of sweet potato as compared to control. Data on soil moisture content and nutrient available status have supported the above result. The positive effect of zeolites is considered due to twin benefits, viz. soil water and nutrient retention. In India, successful production and application of FAZ will have an impact on improved utilization of fly ash in agriculture and the levels of extractable toxic elements will be significantly decreased in the fly ash due to zeolitization that result in minimizing the environmental pollution due to landfill, ground water pollution, etc.

2 Research Status

Information also available on the possible use of FAZ in agriculture and is widely tested in countries like Australia and China in soils of varied texture and at different rates on the growth of canola, spinach, and wheat. In India, fly ash is added to compost at specific ratio and its performance on changes in soil properties (Ramesh and Chhonkar 2001) as well as in tree species (Ramesh et al. 2008) was studied. Extensive works on fly ash zeolites have been carried out by NEERI and NCL, Pune. Rayalau et al. (2001) synthesized zeolites from different sources of coal and its implications on Zeolite-A production. The zeolites thus produced were also characterized for cystallinity, calcium-binding capacity, and sorption properties. Ohja et al. (2004) synthesized and characterized zeolite from fly ash using hydrothermal treatment and compared the cost factor with that of zeolite available in market. Though many studies on zeolite synthesis have been carried out, no studies so far attempted to produce a neutral pH-based zeolite which only will be useful for land applications. Otherwise the high pH of the zeolite has deleterious effect on soil carbon content and availability of other major and micronutrients, which are highly pH-dependent. Few works on commercial zeolites (Zeolite 4A) were initiated at ICAR-CRIDA to improve the efficiency of water and nutrients in drylands. Good number of studies is initiated by fly ash mission as technology demonstration projects utilizing fly ash under different agro-climatic conditions and soil types and the potential of fly ash for agriculture in India is also extensively reviewed (Alam and Akhtar 2011). The importance of value-added fly ash was also emphasized earlier for better utilization (Jala and Goyal 2006).

Knowledge Gaps Review of works conducted in India and outside revealed that research work on synthesis and field applications of agricultural grade fly ash zeolites in agriculture in general and tuber crops in particular especially in relation to soil physical characteristics and organic carbon content are not available and hence the proposed objectives (other than production of FAZ) are new and assumes a lot of practical significance in the utilization of fly ash in agriculture.

In literature, the beneficial aspects and scenario of fly ash production, utilization, and soil management are reviewed extensively (Ahmaruzzaman 2010; Senapati 2011). Dhadse et al. (2008) quoted that even if 10% cultivators use fly ash @10MT/ha, agricultural sector will require 170 MT FA each year which will fall short of the total fly ash production (112 MT/year). Haque (2013) reported the utilization statistics of fly ash and mentioned its decreased utilization based on 2010–12 statistics. Reports also indicated that low dose of fly ash (2–4%) has shown increase in average plant growth parameters.

Zeolites are crystalline aluminum–silicates with group I and II elements as counter ions. The zeolite yield is highly dependent upon the glass phase (amorphous) of fly ash which again depends on composition and burning efficiency of coal at TPP. Their structure is made up of a framework of $(SiO_4)^{4-}$ and $(AlO_4)^{5-}$ tetrahedral linked to each other at the corners by sharing their oxygen.

Root crops are recognized as "future crops" or "food security crops." Cassava and sweet potato were selected for the present study because these two crops account for more than 30% of total root crops from developing countries because of their economic, commercial, and industrial importance (in starch, sago, animal feed, bioethanol, etc.) and the role played by the crops in improving the livelihood of poor farmers. A report of International Food Policy Research Institute has reported a projected demand of 37% increase for root and tuber crops globally between 1995 and 2020. The worldwide demand for cassava and minor tuber crops is projected to increase by 49% and sweet potato and yams by 30%. Likewise, India needs 1.27 lakh tones of sweet potato by 2016. Similarly, cassava starch finds varied uses in India in food industry for the production of Sago (Sabudhana) and many other industrial applications. Low soil fertility is the single most pervasive constraint to high and sustainable production worldwide and it was reported that cassava yield steadily declines without the use of soil amendments.

Soil organic matter directly influences soil properties due to its effects on soil properties (Wendling et al. 2010). Studies also indicated fly ash can be more beneficial when used in combination with organic additions. Kishor et al. (2010) mentioned the advantage of fly ash in reducing carbon dioxide emission when used in soil instead of lime. Palumbo et al. (2007) studied the fly ash characteristics in relation to carbon sequestration potential and reported that careful selection of fly ash can maximize the potential for carbon sequestration and minimize chances of toxicity when applied to degraded land. Wang and Zhu (2007) tested fly ash with different contents of unburned carbon for humic acid adsorption and investigated the influence of unburned carbon on adsorption. Experiments were conducted to understand the decomposition resistance of ca zeolites and organic matter at high temperatures showed a slight increase in CN ratio of soils. In addition, carbon accumulation in humic fractions as well as degree of humification and aromacity of humic acids increased. This suggests that the presence of zeolites could be beneficial for soil organic matter conservation under global warming (Truc and Yoshida 2011). Soil organic matter labile fractions are used instead of total organic matter as sensitive indicators of change in soil quality (Haynes 2005).

3 Synthesis of Zeolite Types—Lab Scale Studies

Sieved fly ash (250 μm) was used for the synthesis study. Few grams of fly ash were subjected to pre-treatment using conc. HCl with simultaneous heating and stirring actions. Then, the product was repeatedly washed with distilled water to remove the acid traces and oven dried at 100 °C for 12–18 h. This treatment helped to reduce the iron (favorable condition for more zeolitization) and heavy metal contents because of agricultural applications. More research efforts must be put in this line of research as the two most important qualities that a fly ash-based zeolites should possess for use as agricultural grade must be a near-neutral pH and minimum heavy metal content. This could be further tested by periodic sampling of the zeolite applied soils and plants for the important heavy metal content of interest.

Two different methods were attempted in gram-level zeolite synthesis. They are (1) Simple hydrothermal method and (2) alkali fusion followed by hydrothermal method. In the first method, zeolites were synthesized from pre-treated fly ash using three different combinations of mineralizers, viz. NaOH—2 M; KOH—2 M; and equimolar solutions of NaOH and KOH—2 M. In addition, the following synthesis conditions were followed during synthesis: Solid/liquid ratio—1:10; temperature—121 °C; pressure—15 psi; and time—48 h. The autoclaved product was repeatedly washed in distilled water and dried. The final pH of the washed aliquot was found to be 10.6.

In the second method, the pre-treated fly ash was mixed with the mixture of NaOH and KOH, thoroughly ground and heated in muffle furnace and the powdered product was reacted with water and autoclaved for 24 h. This was followed by repeated washing and the finished product was dried at 90 °C for 24 h. XRD analysis revealed that it is predominantly consisted of different zeolites types, viz. zeolite X, analcime, faujasite, chabazite, and sodalite. The pH of synthesized products ranged from 10.5 to 10.9. The XRD results of the synthesized zeolites mixtures using the two methods showed considerable variations especially when compared with the XRD pattern of commercial zeolite 4A which was produced by perfect chemical industrial process (Fig. 1).

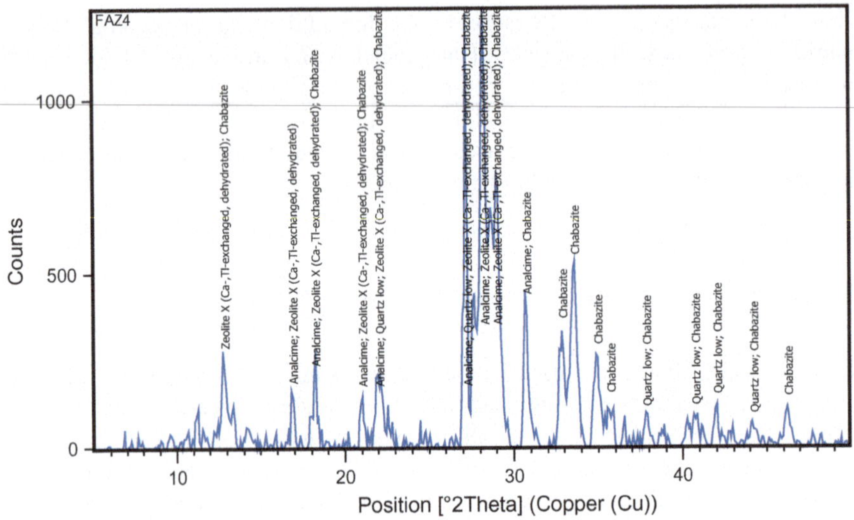

Fig. 1 FAZ synthesized using NaOH and KOH 1 M each (alkali fusion)

4 Characteristics of Bulk Synthesis Fly Ash Zeolites for Field Applications

In each batch, 500 g of pre-treated fly ash was mixed with alkali mixtures (1 M KOH + 1 M NaOH) in standardized proportions and mixed with distilled water with thorough stirring. Then, the contents were taken in a stainless steel vessel and autoclaved for 48 h. After the reaction, the product was washed thoroughly with distilled water to remove the excess sodium contents and the product was dried in an oven. This product had a pH of 9.92. Then, the synthesised zeolites were subjected pH reduction using acetic acid treatment and then repeatedly washed with distilled water. The synthesized pH-treated zeolite mixtures for field application had a pH of 6.68, CEC of 254.13 cmol (p+) kg^{-1} and an available sodium content of 27.18 g kg^{-1}. Analysis of SEM also showed considerable zeolitization of fly ash at different magnitudes (Fig. 2). The mean particle size of synthesized zeolites decreased from 40 μm (corresponds to fly ash) to 25 μm with 100% of the particles fall below 100 μm size).

XRD results of bulk synthesized zeolites show well-characterized and distinct peaks correspond to zeolite X, sodalite, and chabazite. Absence of amorphous-related peaks further showed the efficient conversion of Si and Al in glass phase to zeolites. Thus, the characterization studies including definite crystallization as obtained from SEM confirms the zeolite quality (pH-treated zeolites) with respect to pH and CEC. It could be observed from the images especially the characteristic well-developed formation of zeolites formed from the fly ash particles. Thus, the results of XRD, SEM, and the results that were related to the physicochemical characteristics of the synthesized zeolites correlated well with the high cation-exchange capacity of the product. As mentioned earlier, the purpose of the work is to synthesize agricultural

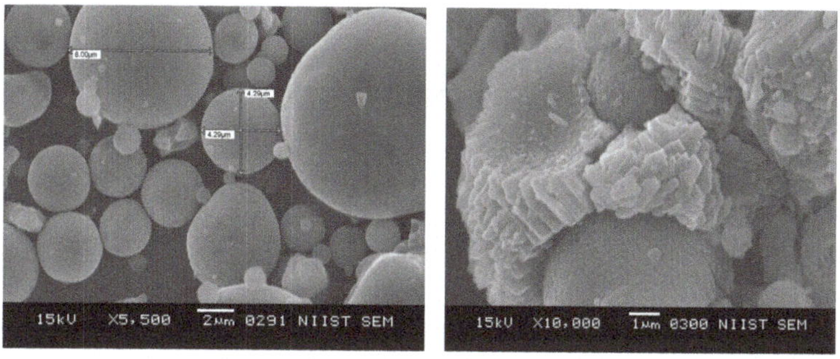

Fig. 2 SEM pictures of fly ash (left) and zeolites synthesized in bulk quantity (right)

grade fly ash zeolite mixtures rather than single mineral of zeolites. Hence, the effect of different concentration of mineralizers and synthesis conditions will further improve the product quality.

5 Soil Application Studies

Initial studies were conducted at ICAR-CTCRI on synthesis, characterization, and evaluation/application of FAZ in sweet potato. Synthesis made using hydrothermal method has shown the following characteristics: Texture—Silt, pH—6.68, CEC-254.1 cmol (p+) kg^{-1}. Soils amended with 1% FAZ by weight had high soil water retention (19.2%, v/v) as compared to control (15.8%). The tuber yield was found to be 57% higher over control (Fig. 3).

In another study on soil and nutrient adsorption properties of dryland Alfisols amended with commercial synthetic zeolite 4A, the changes in chemical properties of different soil–zeolite mixtures was observed under incubated conditions with constant temperature and soil moisture conditions and the adsorption pattern of ammonium in different soil–zeolite mixtures were studied. The results of the study indicated

Fig. 3 View of the pot experiment (left) and tubers obtained in soils amended with FAZ (right)

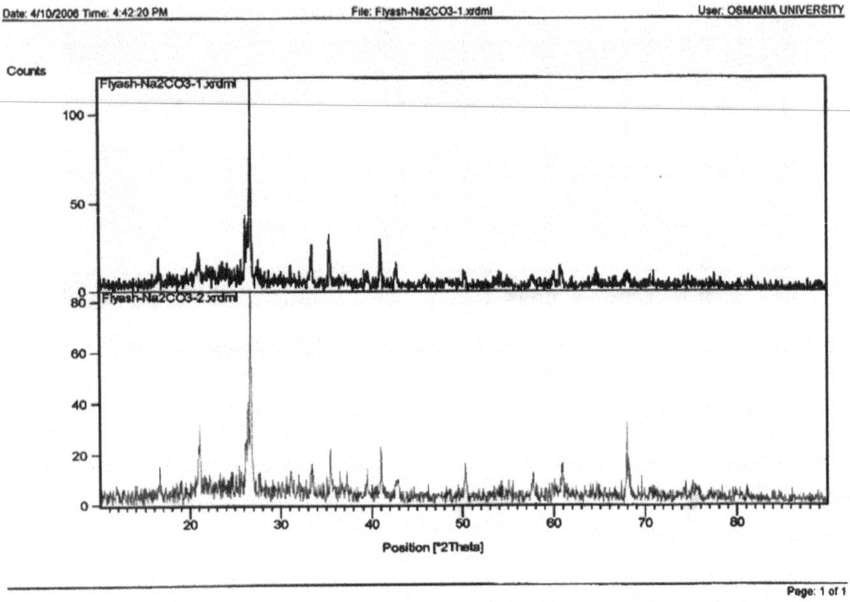

Fig. 4 XRD results of synthesized zeolites using VTPS fly ash

that zeolite 4A was found to possess high CEC (1564 mmol/kg) and water retention properties 4 and 2.5 times as compared to soil at 0.33 and 15.0 bar, respectively.

An experiment was conducted on the synthesis of zeolites using fly ash collected from Vijayawada Thermal Power Station (VTPS) using two different mineralizers, viz. sodium hydroxide and sodium carbonate and the synthesized product was analyzed for the quality and mineral synthesized using XRD. The two Theta values were compared with the commercial 4A zeolite, which has distinct peaks that indicated the high degree of crystallization and product quality while FAZ was found to have inferior quality (Fig. 4). In another field study, zeolite 4A was evaluated under microsite (pit) conditions based on the results of laboratory studies for a dryland horticultural crop, mango. In order to improve the efficiency of zeolites, two different methods of application, viz. uniform mixing with soil and band application (15 cm thick band) in the periphery of pit were tried. Commonly available low-cost materials having similar moisture and nutrient benefits such as Fuller's earth, bentonite, and black soil were also included in the study to evaluate and compare for its efficacy. Zeolites and other conditioners were applied at 1% rate (on dry weight basis of soil weight) per pit. Few soil physical observations, viz. profile soil moisture (0–0.4 m depth) and penetration resistance were recorded.

References

Alam, J., & Akhtar, M. N. (2011). Fly ash utilization in different sectors in Indian scenario. *International Journal of Trend in Research and Development, 1*(1), 1–13.

Ahmaruzzaman. (2010). A review on the utilization of fly ash. *Progress in Energy and Combustion Science, 36*(3), 327–363.

Dhadse, S., Kumari, P., Bhagia, L. J. (2008). Fly ash characterization, utilization and government initiative in India—A review. *Journal of Scientific & Industrial Research, 67*, 11–18.

Haque, E. (2013). Indian fly ash production and consumption scenario. *International Journal of Waste Resources, 3*(1), 22–25.

Haynes, R. J. (2005). Labile organic matter fractions as central components of the quality of agricultural soils. *Advances in Agronomy, 85*, 221–268.

Jala, S., & Goyal, D. (2006, June). Fly ash as a soil ameliorant for improving crop production—A review. *Bioresource Technology, 97*(9), 1136–1147. Epub 2004 November 11.

Ojha, K., Pradhan, N. C., & Samanta, A. N. (2004). Zeolite from fly ash: Synthesis and characterization. *Bulletin of Material Science, 27*, 555–564.

Palumbo, A. V., Tarver, J. R., Fagan, L. A., McNeily, M. S., Ruther, R., Fisher, L. S., et al. (2007). Comparing metal leaching and toxicity from high pH, low pH and high ammonia fly ash. *Fuel, 86*(10–11), 1623–1630.

Kishor, P., Ghosh, A. K., & Kumar, D. (2010). Use of fly ash in agriculture: A way to improve soil fertility and its productivity. Asian *Journal of Advances in Agricultural Research, 4*(1), 1–14.

Ramesh, V., Korwar, G. R., Mandal, Uttam Kumar, Prasad, J. V. N. S., Sharma, K. L., Ramakrishna, Y. S., et al. (2008). Influence of fly ash mixtures on early tree growth and physicochemical properties of soil in semi-arid tropical Alfisols. *Agroforestry Systems, 73*(1), 13–22. (Netherlands).

Ramesh, V., & Chhonkar, P. K. (2001). Chemical characteristics of an acid sulphate soil from Kerala amended with lime and fly ash. *Journal of Indian Society of Soil Science, 49*(4), 719–726.

Rayalu, S., Udhoji, J. S., Munshi, K. N., & Hasan, M. Z. (2001). Highly crystalline zeolite-A from fly ash of bituminous and lignite coal combustion. *Journal of Hazardous Materials, 88*, 107–121.

Senapati, M. R. (2011). Fly ash from thermal power plants—Waste management and overview. *Current Science, 100*(25), 1791.

Truc, M. T., & Yoshida, M. (2011). Effect of zeolite on the decomposition resistance of organic matter in tropical soils under global warming. *International Scholarly and Scientific Research & Innovation, 5*(11), 664–668.

Wang, S., & Zhu, Z. H. (2007). Humic acid adsorption on fly ash and its derived unburned carbon. *Journal of Colloid and Interface Science, 315*(1), 41–46.

Wendling, B., Jucksch, I., Mendonca, E. S., & Alvarenga, R. C. (2010). Organic-matter pools of soil under pines and annual cultures. *Communications in Soil Science and Plant Analysis, 41*, 1707–1722.

Pesticidal Activity and Future Scenario of Fly ash Dust and Fly ash-Based Herbal Pesticides in Agriculture, Household, Poultry and Grains in Storage

Y. Hariprasad, C. Kathirvelu and P. Narayanasamy

Abstract Results showed that the fly ash as dust could be an active control factor in checking the infestation of crop pests and also safe to different non-target organisms like natural enemies, beneficial insects hovering in and around agro-ecosystems. Application of fly ash @ 40 kg/ha in rice had caused maximum mortality of 73.3% in leaf caterpillar followed by grasshopper (71.0%), horned caterpillar (68.8%). Nine fly ash-based herbal insecticides, viz. neem (*Azadiracta indicia* A. Juss.), Indian Privet (*Vitex negundo* Linn.), *Eucalyptus* (*Eucalyptus globulus* Labill.), Pongam (*Pongamia pinnata* Linn.), Turmeric (*Curcuma longa* Linn.), Chilli (*Capsicum annuum* Linn.), Sweet flag (*Acorus calamus* Linn.), Ginger (*Zingiber officinale* Roscoe) and Adathoda (*Adhatoda vasica* Linn.) were synthesised using fly ash as a carrier. The eventual fly ash-based herbal pesticides were evaluated against key pests of crops. In vegetables, the fly ash-based herbal pesticides manifested better results against major pests of bhendi and brinjal. Results showed that the combination of treatment fly ash + square stalked vein 10% D acted very effective against mandibulate termite soldier with maximum mortality followed by fly ash + French basil 10% D. The overall mean efficacy of lignite fly ash against housefly, a menace around poultry yard was observed at 83.33% mortality of maggots. These new class of herbal pesticides had also exhibited good control of insect pests, namely *Tribolium castaneum* Herbst, *Callosobruchus chinensis* Linn., attacking stored product.

Keywords Herbal pesticides · Formulations · Environment · Pesticidal activity

Y. Hariprasad (✉) · C. Kathirvelu · P. Narayanasamy
Department of Entomology, Faculty of Agriculture, Annamalai University, Annamalainagar, Tamil Nadu 608 002, India
e-mail: yagamurhari@yahoo.com

C. Kathirvelu
e-mail: ckathirveluau@gmail.com

P. Narayanasamy
e-mail: drpnsamy@yahoo.com

S. K. Ghosh and V. Kumar (eds.), *Circular Economy and Fly Ash Management*,
https://doi.org/10.1007/978-981-15-0014-5_5

1 Introduction

The annual production of fly ash in India was about 169.25 million tonnes (MTs), and utilization was 107.10 MTs and about 63 MTs remained unutilized during the year 2016–2017 (CEA Report 2017). This huge volume of fly ash requires large areas of lands in the form of ash ponds for dumping which may lead to encroachment on agricultural land. Such a huge volume possesses a challenging threat, in the form of usage of land, health and environmental hazards. Of late, control of the pest problems in agriculture with natural substrates especially plant products has been emphasized very much. Similarly, lignite fly ash (LFA), a by-product obtained as waste by combustion of lignite coal in the thermal stations of India, has become an environmental hazard by its voluminous heaping in open spaces and the surrounding environment gets abundantly polluted. Among various ingredients of the ash, silica (as SiO_2 60%) which strengthens use of the fly ash has been detected to be insecticidal against various pest problems in variety of crops (Narayanasamy and Gnanakumar Daniel 1989; Arputhasankari and Narayanasamy 2007b).

2 Generation and Disposal of Fly ash in India

In India, maximum utilization of fly ash was recorded, during the year 2016–17, to a tune of 40.58% of total fly ash utilized in the cement sector, followed by 14.91% in making bricks and tiles, 11.88% in ash dyke raising, 11.78% in mine filling, 11.03% in reclamation of low lying area, 6.19% in roads and embankments, 1.92% in agriculture, 0.76% in concrete and 7.98% in others, etc. (CEA Report 2017). Scientific disposal and management of the fly ash is still the most important problem encountered by coal-based thermal power plants. Many technologies have been developed for effective management of fly ash through Fly ash Utilization Programme operated by Fly Ash Mission under Ministry of Science and Technology, Government of India, New Delhi.

3 Utilization of Fly ash

Fly ash is characterized by its constitution with various chemical elements, viz. silica, alumina, iron and copper, having low specific gravity, lack of plasticity and more of pozzolanic action. It has been found beyond doubt, to be useful in preparing various products such as house-building bricks, Portland pozzolanic cement, hollow concrete blocks, lime–fly ash aggregate mixture. The fly ash is also put into use as structural fills, quarry fills, and in strengthening embankments, dikes and dams. It has been found suitable as sub-base course and base course in the construction of highway pavement. Further, the fly ash use in agricultural fields has shown multivarious sectors both in the cop production and protection.

4 Agricultural Application of Fly ash

In agriculture, the fly ash has been shown to be serving as manure providing certain micro-nutrients for the crop plants. Other features like conditioning of the soil through correction of deficient nutrients, prevention of soil erosion and induction of plant resistance against insect pests are also anchored by the fly ash. During the last decade, variety of utilities of the fly ash has been brought to light.

5 Control of Insect Pests with Fly Ash

5.1 Fly ash Against Field Crop Pests

Narayanasamy (1999) reported for the first time insecticidal property of the fly ash. In array of crop pests like, leaf folder, stem borer, grasshopper and sucking pests like brown planthopper (BPH), green leafhopper (GLH), earhead bug, beetles were controlled significantly compared to the commercial insecticide, BHC 10% dust used as check. A dosage of 40 kg per hectare was found optimum for control of the pest problems. Application of the ash at 30 and 50 days after treatment (DAT) in rice was found effective.

Insecticides like BHC 10% dust, Malathion 50% WDP, BHC 50% WDP and Carbofuran 3% WDG were prepared with LFA as the main carrier following ISI code available for such pesticides. These pesticides efficiently checked the populations of leaf folder, yellow hairy caterpillar, small grasshopper's earhead bugs, GLH and BPH (Narayanasamy 1994).

Results showed that the fly ash killed the pests significantly. The mortality rate was found to be highly pronounced in the leaf folder larvae with 73.33%, followed by grasshopper (71.10%), yellow hairy caterpillar (68.88%) and yellow rice borer (46.66%). The spiny beetles were controlled to the least (39.99%).

The results revealed that the susceptibility of the pests to the fly ash dust varied significantly among themselves. Highest mortality of 80% was recorded in earhead bug and a lowest of 38.88% with brown bug.

5.2 Vegetable Crop Pests

In a laboratory study, Selvanarayanan et al. (1997) reported that the highest dose (125 mg) of fly ash had caused minimum mortality (22.14%), while low doses took long time to cause larval mortality. Gnanakumar Daniel (1989) made a maiden research on the use of fly ash against variety of insect pests concluding insecticidal action of the ash.

In the case of *Crocidolomia binotalis* Fab., it was observed that all the doses of fly ash in general had caused 100% mortality. But when overall performance was considered, 500 mg LFA was the most effective treatment followed by 250, 750 and 1000 mg fly ash doses (Table 1) Gnanakumar Daniel 1989; Narayanasamy and Gnanakumar Daniel 1989; Narayanasamy 1994).

Nambirajan (1999) reported that the insect pests (*Earias vittella* Fab., *Helicoverpa armigera* Hub.) showed their susceptibility to lignite fly ash treatment with maximum mortality. Coal–lignite fly ash (CLFA) also revealed its efficacy against pests like *E. vittella, Sylepta derogate* Fab., *Xanthodes graellsi* Feis., *H. armigera* and *S. litura* with significant mortality. Among them, pests like *E. vittella, X. graellsi* and *H. armigera* were found highly susceptible.

LFA and CLFA were found equally efficient against the brinjal pest, namely fruit borer, and *Leucinodes orbonalis* Guen. found with maximum susceptibility to the LFA and CFA treatments.

Table 1 Insecticidal activity of different doses of lignite fly ash against the fifth instar larvae of *Crocidolomia binotalis* Zeller in Cauliflower in laboratory

Dose of fly ash (mg/leaf)	Per cent mortality at different hours after treatment			Mean mortality (%)
	24	48	72	
250	44.00 (41.55)	45.00 (42.13)	100.00 (90.000)	63.00 (52.54)[a]
500	44.00 (41.55)	56.67 (48.83)	100.00 (90.00)	66.89 (54.87)[a]
750	96.00 (78.46)	20.00 (26.57)	0.00 (5.74)	38.67 (38.45)[b]
1000	96.00 (78.46)	20.00 (26.57)	0.00 (5.74)	38.67 (38.45)[b]
Untreated control	0.00 (5.74)	12.00 (20.27)	18.00 (24.35)	100.00 (18.43)[c]
SED				3.84
CD ($p = 0.05$)				8.14

Figure in parentheses are arcsine transformed values
Values mean of five replications
Means followed by a common alphabet are not significantly different (at 5% level by DMRT)
Selvanarayanan et al. (1997)

Dusting of LFA and CLFA over the body of the test insects showed that all the insects were highly susceptible in which mortality ranged between 40 and 60%. It was significant to note that the early larval insects were more susceptible to the treatments than the advanced stages. Among the pests tested, S. *litura* larvae showed highest mortality of 55.39% among *H. armigera, L. orbonalis* and *E. vittella.* Between the fly ashes, lignite fly ash was found to be the best against the pest tested as to the CLFA.

Results showed that the treated pupae of *E. vittella,* succumbed with cent per cent mortality while other insects were dead by 40–50%. It was interesting to note that in those cases from which adult moths emerged, adult malformation of varying range was witnessed in the adult stage. Both LFA and CLFA expressed their physical impact on the pupae through pupal mortality, and adult malformations were also noticed (Nambirajan 1999).

The leafhoppers of bhendi succumbed to death due to feeding of the fly ashed diet having 2.5% concentration. Both LFA and CLFA were equally effective against the hopper. In contrast, pests like *E. vittella, L. orbonalis* and *H. armigera* were found to thrive on LFA treated plant parts and developed successfully into normal adults.

The caterpillars poisoned due to LFA showed an almost identical symptom regardless of the species. Dissection of the insect gut of the dead insects revealed that the ash particles in general settled either in the foregut or midgut in various insects. The feature could have hampered the digestion of food and blocked the food pathway, thus paralyzing the activity of the entire gut. Chemical elements like silica, alumina and zinc present in the fly ash could have played role in this regard and silica, in particular, with it's high content of about 58% as reported by Raghupathy (1988) would have taken a lead role.

It was therefore concluded that LFA dusts through various mechanisms causing insect mortality could be involved as dust insecticide for most of the surface feeding insects.

6 Store Grain Pests

Muthukumaran et al. (2001) in a preliminary experiment reported that the mortality of pulse beetles increased steadily as the day progressed, all the insects in various pulses died on fourth day after treatment.

Storage pests, namely *C. chinensis* and *T. castaneum,* were checked effectively with fly ash-based herbal pesticides, namely FA + neem, FA + vitex, FA + eucalyptus, FA + pongam, FA + turmeric, FA + chilli, FA + acorus, FA + ginger and FA + adathoda during and up to 90 days of storage (Tables 2 and 3) (Coal India Project Report 2011).

Table 2 Population development of *Tribolium castaneum* as influenced by fly ash-based herbal pesticides in store godown

S. No.	Treatments (@ 2%)	Number of adult beetles survived/½ kg of seeds days after application						Overall Mean
		15	30	45	60	75	90	
1	FA + Neem 30% D	0.00	15.33	30.00	40.00	50.00	80.00	35.89[d]
2	FA + Vitex 30% D	3.33	12.00	25.00	40.00	58.00	68.00	34.39[c]
3	FA + Eucalyptus 30% D	8.00	14.00	30.00	46.00	65.00	73.00	39.33[f]
4	FA + Pongam 30% D	4.00	10.33	28.33	42.00	62.00	70.00	36.11[e]
5	FA + Turmeric 30% D	5.00	11.00	34.67	54.00	69.00	77.33	41.83[g]
6	FA + Chilli 30% D	2.00	15.33	32.33	50.00	73.00	81.00	42.28[h]
7	FA + Acorus 30% D	0.00	8.00	15.00	25.00	38.00	45.00	21.83[a]
8	FA + Ginger 30% D	7.67	16.67	36.00	58.00	78.00	83.67	46.67[i]
9	FA + Adathoda 30% D	0.00	9.00	20.67	30.00	45.33	52.00	26.16[b]
10	Untreated control	9.00	25.00	40.00	60.00	85.00	120.00	56.50[j]
CD ($p = 0.05$)								0.599
SED								0.298

Mean of three replications
Means with different alphabet do vary significantly according to DMRT
Coal India Project Report (2011)

7 Household Pests

7.1 House Fly

Indumathi (2009) made a maiden research on tackling house fly with fly ash and reported that four kinds of ashes, namely lignite fly ash (LFA), rice husk ash, coir pith ash and cow dung ash, were tested for their bio-efficacy against the housefly maggots at 3, 5, 10 and 15% concentrations.

Results showed that overall mean efficacy in insecticidal activity of lignite fly ash (LFA) against housefly. Among all the dosages tested, five gram incited maximum level of mortality (83.33%) followed by rice husk ash (RHA) (74.99%) and coir

Table 3 Population development of *Callosobruchus chinensis* by fly ash-based herbal pesticides in godown +

S.No	Treatments (@ 2%)	Number of adult beetles survived/½ Kg of seeds days after application						Overall mean
		15	30	45	60	75	90	
1	FA + Neem 30% D	15.0	30.0	49.0	65.0	73.0	87.0	53.17[g]
2	FA + Vitex 30% D	7.0	24.0	38.0	58.0	70.0	80.0	46.17[c]
3	FA + Eucalyptus 30% D	12.0	30.0	43.0	62.0	70.0	82.0	49.83[d]
4	FA + Pongam 30% D	8.0	28.0	40.0	60.0	78.0	88.0	50.33[e]
5	FA + Turmeric 30% D	9.0	32.0	45.0	64.0	74.0	85.0	51.5[f]
6	FA + Chilli 30% D	9.0	35.0	48.0	70.0	85.0	93.0	56.67[h]
7	FA + Acorus 30% D	2.0	20.0	35.0	48.0	57.0	69.0	38.50[a]
8	FA + Ginger 30% D	7.0	40.0	50.0	75.0	93.0	98.0	60.50[i]
9	FA + Adathoda 30% D	5.0	25.0	40.0	52.0	63.0	71.0	42.67[b]
10	Untreated control	22.0	45.0	57.0	80.0	100.0	130.0	72.33[j]
CD ($p = 0.05$)								0.692
SED								0.344

Mean of three replications
Means with different alphabet do vary significantly according to DMRT
Coal India Project Report (2011)

pith ash (70%). While least insecticidal activity was witnessed with cow dung ash (36.66%) followed by coir pith ash (43.33%) at 3% concentrations, the treatment lignite fly ash 5 g caused highest per cent mortality of maggots (90.00%) and malformation of puparia (76.66%) followed by rice husk ash (83.33 and 66.66%) and coir pith ash (80.00 and 60.00%). In contrast, the cow dung ash (43.33 and 30.00%) caused least per cent mortality and malformation of puparia at 3% concentrations (Table 4).

7.2 Termites

Arputhasankari and Narayanasamy (2007a) attempted to tackle the wood eating pest, termites with fly ash and reported that the mixture of fly ash and moist soil (1:1 ratio) had caused 24.09% mortality of the termite and prohibited free movement of other surviving insects. In the treatment of rotten wood with fly ash paste, a maximum of

Table 4 Bio-efficacy of certain ashes against various developmental stages of the housefly maggots (*Musca domestica*)

S. No.	Treatments	% Maggot mortality	% Malformed puparium	Over all mean efficacy
1	Lignite fly ash dust @ 3 g	60.00 (51.14)^{jk}	53.33 (46.92)^{i}	56.66 (49.03)^{J}
2	Lignite fly ash dust @ 5 g	90.00 (76.22)^{a}	76.66 (62.71)^{a}	83.33 (69.46)^{a}
3	Lignite fly ash dust @ 10 g	73.33 (59.2 1)^{f}	60.00 (51.14)^{d}	66.66 (55.17)^{f}
4	Lignite fly ash dust @ 15 g	80.00 (64.63)^{c}	66.66 (56.07)^{b}	73.33 (60.35)^{c}
5	Rice husk ash dust @ 3 g	53.33 (46.92)^{m}	46.66 (43.07)^{k}	49.99 (44.99)^{l}
6	Rice husk ash dust @ 5 g	83.33 (66.63)^{b}	66.66 (56.07)^{bc}	74.99 (61.35)^{b}
7	Rice husk ash dust @ 10 g	66.66 (49.92)i	56.66 (48.92)^{gh}	61.66 (49.41)^{l}
8	Rice husk ash Dust@ 15 g	80.00 (63.92)^{cd}	60.00 (51.14)^{de}	70.00 (57.53)^{d}
9	Coir pith ash dust @ 3 g	46.67 (46.92)^{o}	40.00 (38.85)^{mn}	43.33 (42.52)^{n}
10	Coir pith ash dust @ 5 g	80.00 (70.37)^{de}	60.00 (51.14)^{ef}	70.00 (60.75)^{de}
11	Coir pith ash dust @ 10 g	60.00 (51.14)j	46.66 (43.07)^{kl}	53.33 (47.10)^{k}
12	Coir pith ash dust @ 15 g	70.00 (56.78)^{g}	56.66 (46.92)^{g}	63.33 (51.85)^{g}
13	Cow dung ash dust @ 3 g	43.33 (41.07)^{p}	30.00 (33.21)^{p}	36.66 (37.14)^{p}
14	Cow dung ash dust @ 5 g	70.00 (56.78)^{gh}	50.00 (45.00)^{j}	60.00 (50.89)^{i}
15	Cow dung ash dust @ 10 g	50.00 (45.00)^{n}	36.66 (34.06)^{o}	43.33 (39.53)^{no}
16	Cow dung ash dust @ 15 g	56.66 (48.92)^{l}	40.00 (38.85)^{m}	48.33 (43.88)^{m}
17	Untreated control	0.000 (4.504)	0.000 (4.054)	0.000 (4.504)
SED		9.28	11.00	10.14
CD		18.86	22.37	20.61

Values in parentheses are arcsine transformed values
Each value is a mean of three replications used @ 10 maggots/replication
Means with different alphabet differ significantly according to DMRT
Indumathi (2009)

25.75% mortality of termites was recorded. In the rotten wood treated with fly ash, a maximum of 24.93% mortality of termite was noticed. Topical application of fly ash on the termites had affected 27.27% mortality. Among the four modes of fly ash treatments, topical application of fly ash on the termites was found superior recording 27.27% mortality followed by fly ash paste treatment, fly ash dusted wood and mixing of fly ash with moist soil with 25.75, 24.93 and 24.09% mortality, respectively.

Selvi (2008) developed three herbal insecticides with fly ash as a carrier and evaluated them against termites. She reported that worker termites were killed effectively by dusting with treatment FA + square stalked vein 10% D with 93.33% followed by FA + French basil 10% D (90.00%) and FA + black pepper 10% D (86.67%).

The treatment FA + square stalked vein 10% dust acted very effectively against mandibulate termite soldier with maximum mortality followed by FA + French basil 10% D. Fly ash + black pepper 10% D treatment had least effect on the insect. The data revealed that the treatment FA + square stalked vein 10% D had incited highest mortality of the mandibulate soldier with 90.00 and 73.33% mortality at 24 and 48 h, respectively.

Among all the treatments, FA + square stalked vein 10% D acted very effectively against nasute soldier with 80.00 and 63.33% followed by FA + French basil 70.00 and 60.00% at 24 and 48 h after treatment. The minimum efficacy was witnessed with FA + Black pepper 10% D treatment with 36.67 and 46.67% at 24 and 48 h after treatment.

FA + Square stalked vein 10% D had controlled effectively at 5% dosage at 24 and 48 h after treatment while FA + Black pepper caused least mortality at 1% at 24 and 48 h after treatment. All the treatments were on par with each other and FA + Square stalked vein 10% D, in particular, caused maximum mortality of worker termites followed by mandibulate and nasute soldier termites (Table 5) (Selvi 2008).

She further found that the fly ash pasted wood did cause maximum mortality (32.51%) of the mandibulate soldiers followed by fly ash soil mixture (31.80%). Topical application of fly ash excelled with 33.21% mortality of the termites compared to the treated wooden material (31.09%).

Among the four methods of application, highest per cent mortality (33.21%) of the mandibulate soldier termites was recorded with topical application followed by fly ash paste application (32.51%).

Immediately after releasing the termites at the central point, their behavioural features were tackled. As soon as the termites were released, they moved towards the fly ash heap but could not move for a short span of time before going away from the fly ash dusts finally. Those termites which went towards the treated wood did move easily without feeding. Their entire movements were directionless around the vicinity of the treated food. At last, all the termites reached the spot where the untreated food was placed and started feeding on the inner part of the rotten wood. This shows that deterrent nature of the fly ash had driven the termites.

When the performance of four methods of fly ash application against the workers and soldier termites was compared, topical application of fly ash on the termites was found better with 26.56% mortality followed by fly ash pasted rotten wood (25.75%). Likewise, there was a maximum of 33.14% mortality of the mandibulate-type soldiers

Table 5 Bio-efficacy of fly ash-based herbal pesticides against termite forms (*Hypotermes obscuriceps* Wasmann) in laboratory

S. No.	Treatments	Dosage in grams/petriplate	Per cent mean mortality of termite forms after hours of treatment								
			Workers			Mandibulate soldier			Nasute soldier		
			24	48	Overall mean efficacy	24	48	Over all mean efficacy	24	48	Overall mean efficacy
1	FA + French basil 10% D	1	56.67	70.00	63.34	46.67	63.33	55.00	40.00	53.33	46.67
			(48.85)	(56.79)	(52.82)	(43.08)	(52.75)	(47.92)	(39.22)	(46.92)	(43.07)
2	FA + French basil 10% D	3	63.33	79.69	71.51	56.67	76.67	66.67	53.33	60.00	56.67
			(52.71)	(67.11)	(59.91)	(48.85)	(61.12)	(54.98)	(46.91)	(50.77)	(48.84)
3	FA + French basil 10% D	5	70.00	90.00	80.00	63.33	83.33	73.33	60.00	70.00	65.00
			(56.79)	(71.57)	(64.18)	(52.71)	(65.88)	(59.29)	(50.77)	(56.79)	(53.78)
4	FA + Square stalked vein 10% D	1	60.00	73.33	66.67	53.33	66.67	60.00	50.00	66.67	58.33
			(50.77)	(58.94)	(54.86)	(46.91)	(54.76)	(50.83)	(45.00)	(54.79)	(49.89)
5	FA + Square stalked vein 10% D	3	70.00	80.00	75.00	67.67	66.67	66.67	56.67	70.00	63.33
			(56.79)	(63.45)	(60.12)	(54.75)	(54.76)	(54.75)	(48.83)	(56.79)	(52.81)
6	FA + Square stalked vein 10% D	5	76.67	93.33	85.00	73.33	90.00	81.66	63.33	80.00	71.66
			(81.14)	(75.11)	(78.13)	(58.89)	(71.62)	(65.25)	(52.73)	(63.47)	(58.10)
7	FA-Black pepper 10% D	1	50.00	63.33	56.67	36.67	50.00	43.33	36.67	46.67	41.67
			(45.00)	(52.73)	(48.87)	(37.28)	(45.0)	(41.14)	(37.26)	(45.10)	(41.18)
8	FA-Black pepper 10% D	3	56.67	0.73.33	65.00	43.33	56.67	50.00	43.33	53.33	48.33
			(48.85)	(58.94)	(53.89)	(41.15)	(48.83)	(44.99)	(41.16)	(46.89)	(44.02)

(continued)

Table 5 (continued)

S. No.	Treatments	Dosage in grams/petriplate	Per cent mean mortality of termite forms after hours of treatment								
			Workers			Mandibulate soldier			Nasute soldier		
			24	48	Overall mean efficacy	24	48	Over all mean efficacy	24	48	Overall mean efficacy
9	FA - Black pepper 10% D	5	63.33	86.67	75.00	53.33	70.00	61.66	50.00	60.00	55.00
			(52.73)	(68.59)	(60.66)	(46.89)	(56.79)	(51.84)	(45.00)	(50.77)	(47.88)
10	Control		0	0	0	0	0	0	0	0	0
	SED		0.67	1.05	0.51	0.93	1.12	0.76	0.79	1.60	0.58
	CD p(0.05)		1.36	2.12	1.02	1.87	2.27	1.54	1.59	3.22	1.16

*Values in parentheses are arcsine transformed
*Each value is a mean of three replications used @ 10 insects/replication
*Means with different alphabet differ significantly according to DMRT
Selvi (2008)

recorded with topical application of fly ash, followed by fly ash pasted wood with 32.11%. This is mainly due to the phenomenon of adsorptive and abrasive nature of the fly ash, which had caused mortality of the termites.

Studies showed significant insecticidal activity of fly ash against worker termites. Among all the dosages tested 5 g had incited maximum level of mortality 13.67 and 16.33%, respectively, at 6 and 12 h followed by 3 g tested with 12.00 and 14.33% and one gram tested with 9.0% sand 12.00% at 6 and 12 h after treatment.

Results showed that soldier termite form suffered least with one gram dosage tested with 10.00 and 11.00% mortality, respectively, at 6 and 12 h followed by 3 g dosage tested with 12.00 and 14.00% and 5 g dosage tested with 12.67 and 14.67% at 6 and 12 h after treatment. Based on overall mean efficacy, maximum mortality was obtained with 5 g dosage followed by 3 and 1 g against worker termites.

8 Fly ash-Based Herbal Pesticides

8.1 Synthesis and Evaluation of Herbal Pesticides Against Pests

Arputhasankari and Narayanasamy (2007b) synthesized fly ash-based herbal insecticides and found that topical application of FA + Turmeric 10% D showed it to be potential against sucking pests of brinjal like nymph and adult of lace wing bug (*Urentius hystricellus* Rich.), aphids (*Myzus persicae* Sulz.) and mealybug (*Coccidohystrix insoiltus* Green).

The lace wing bug nymphs were also controlled by the treatments, FA + Neem Seed Kernel (NSK) 10% D and FA + Chilli 10% D. While the treatments FA + Eucalyptus 10% D, FA + Vitex 10% D and FA + Chilli 10% D were highly pesticidal against the lace wing bug nymphs. Similarly, lace wing bug adult was dealt by the treatments FA + Ocimum 10% D and FA + Chilli 10% D and FA + Vitex 10% D and FA + Pepper 10% D.

Brinjal aphid *(M. persicae)* was also managed by the treatments, FA + Turmeric 10% D (29.63%) and FA + Acorus 10% D (28.88%), and there was no significant difference between the other treatments. The treatment, FA + Acorus 10% D also behaved alike FA + Turmeric 10% D in controlling the brinjal mealybug (28.12%) followed by FA + NSK 10% D and FA + Chilli 10% D.

Most of the brinjal pests including the defoliator and sucking pests the most effective treatment was that of FA + Turmeric 10% D while the treatment of the insect pests, namely *Epilachna* (grub and adult), lace wing bug (nymph and adult), aphid and mealybug showed their susceptibility with 30.36% (both grub and adult), 36.27% (nymph) and 34.25% (adult), 29.63 and 28.12% mortality, respectively.

Results showed that second instar larvae of *Spodoptera litura* were killed effectively by the treatment FA + Ocimum 10% D (32.51%) followed by FA + NSK 10% D, FA + Eucalyptus 10% D and FA + Vitex 10% D. However, the third instar larvae of *Spodoptera* were effectively controlled by the treatment FA + Pepper 10% D, with 32.51% mortality. It was interesting to note that some of the larvae were induced with physical malformations, thus making them totally unfit for life activities.

The treatment FA + Chilli 10% D showed more effect in controlling the bhendi semilooper with 39.23% than FA + NSK 10% (35.26%) and FA + Turmeric 10% D (35.26%).

The treatment FA + Acorus 10% D acted effectively against the fruit borer with 39.23% mortality followed by FA + Turmeric 10% D (35.26%). The treatments FA + Chilli 10% and FA + Acorus 10% D caused highest control of the leaf feeders, bhendi semilooper and fruit borer. Bhendi leafhopper showed their innate susceptibility to the treatment FA + Chilli 10% D with 37.26% mortality, followed by FA + Pepper 10% D (36.60%). The treatments FA + Acorus 10% D and FA + Turmeric 10% D recorded higher per cent mortality (33.21%) of red cotton bug followed by FA + NSK 10% D, FA + Vitex 10% D and FA + Chilli 10% D with 31.09, 31.09 and 31.09%, respectively. Dusky cotton bug was controlled by the treatment FA + NSK 10% D with 31.09% mortality. Among the eight treatments, FA + Turmeric 10% D had incited maximum mortality (34.04%) of red spider mite followed by FA + NSK 10% D and FA + Eucalyptus 10% D with 33.21% and 33.00%, respectively.

From the above, it is inferred that not only fly ash alone but also fly ash herbal pesticides were found effective.

9 Safety of Fly ash to Non-target Organisms

In a field trial, it was found that all the fly ash treatments safe to all the natural enemy fauna in the field (Vijayakumar and Narayanasamy 2001).

It was recorded that the enhanced doses of fly ash irrespective of its sizes were directly proportional to the increase in the population of the spider, *Argiope catenulata* Dole. Predaceous insects like cricket *Conocephalus* sp. damsel fly, *Agriocnemis pygmaea* Ramb. and a larval parasite *Isotima javensis* Roh. survived more in numbers in fly ash treatments than the control. Occurrence of *Zoophthora radicans* Bref. and Batko, a potential fungal pathogen mycosing rice leaf folder was at maximum infection range found with fly ash treatments, and further, the fly ash did not, however, affect its virulence in field conditions.

It is inferred that in general all the fly ash doses were found to protect spiders, crickets, damsel flies, larval parasitoids including fungal infection.

Regarding other target mammals, it was observed that albino rat, frog and fish did not show any symptom of poisoning due to the fly ash treatments. Various biochemical parameters like initial and final body included the termination of the experiment, contents of blood glucose, serum protein haemoglobin including wet weight of body organs remained unaffected in the animals. Wet weights of the important organs like heart, liver, lungs and kidney of the treated animals were similar with that of the control. The treated animals were, however, found to be quite normal. All the fly ash treatments were found safe without any kind of effect on the earthworms too.

Fly ash thus was found absolutely safe at varying dosages to animals, like albino rats, frogs, fishes, earthworms, while in the case of silkworms and honeybees. There was deterrence of the fly ash to the growth of the silkworm larva and inhibition of colony growth in honeybees. It was also safe to all the beneficial fauna which survive naturally on the insect pests.

10 Conclusion

From the above, it is concluded that the fly ash could be a dust insecticide and also an active carrier material in the formulations of herbal insecticides. By virtue of its application covering wide range of crops in plains and hillocks, major quantity of the fly ash heaped at the power plants could be disposed of substantially. Moreover, such use of fly ash dust provides generation of revenue to the thermal plants. Importantly, this may in the long run lead to replacement of the dust insecticides used extensively now leading to decline in the input of the chemical dusts in the agricultural lands. Further, the upgraded fly ash products like fly ash herbal pesticides will also be worth using.

It is therefore everyone's concern to care for a healthy environment and agriculture by way of promoting the fly ash in pest control strategy.

Ackowledgements The authors are thankful to the Ministry of Coal, Government of India; Department of Science and Technology, Government of India; and Tamil Nadu State Council for Science and Technology, Government of Tamil Nadu for their financial support for researches on fly ash. Further, we are grateful to the authorities of Annamalai University for necessary permission in the conduct of the project and for providing infrastructure facilities for the successful conduct and completion of the project on time.

References

Arputhasankari, S., & Narayanasamy, P. (2007a). Flyash—A termiticide. *Pestology, 30*(8), 39–42.

Arputhasankari, S., & Narayanasamy, P. (2007b). Bio-efficacy of flyash based herbal pesticides against pests of rice and vegetables. *Current Science, 92*(6), 811–816.

CEA Report. (2017). *Report on Flyash Generation at Coal/Lignite Based Thermal Power Stations and its Utilization in the Country for the year 2016–17*. Central Electricity Authority, New Delhi.

Coal India Project Report. (2011). Project Completion Report on *Development and use of Flyash-based pesticides* funded by Ministry of Coal, Government of India, 193p.

Gnanakumar Daniel, A. (1989). *Studies on the value of lignite flyash as dust insecticide against certain insect pests*. M.Sc. (Ag.) thesis, Annamalai University, Annamalainagar, Tamil Nadu, 98p.

Indumathi, K. (2009). *Studies on pestiferous insects in poultry premises of Tamilnadu*. M.Sc. (Ag.) thesis, Annamalai University, Annamlainagar, Tamil Nadu, India, 138p.

Muthukumaran, N., Prabhu, S., Radhakrishnan, V., Hariprasad, Y., & Narayanasamy, P. (2001). Bio assay of flyash against grubs of coconut rhinoceros beetle in farm yard manure medium. In P.

Narayanasamy (Ed.), *Agriclutural Application of Flyash. Proceedings of II National Seminar on Use of Flyash in Agriculture*, Annamalai University, Annamalainagar, Tamil Nadu, India.

Nambirajan, S. G. (1999). *Further studies on the use of Flyash 100% Dust against key pests of certain vegetable crops*. M.Sc. (Ag.) thesis, Annamalai University, Annamalainagar, Tamil Nadu, India, 117p.

Narayanasamy, P. (1994). Final Report of a Project titled, "*Studies on use of lignite flyash as an insecticide and an adjuvant in insecticide formulations*" supported by the Tamil Nadu State Council for Science and Technology, Chennai, Tamil Nadu, India, 99 p.

Narayanasamy, P. (1999). Flyash as pesticide in agriculture. In A. Sridharan, N. S. Pandian, & Vimalkumar (Eds.), *Flyash Characterization and Its Geotechnical Applications. Proceedings of National Seminar on Flyash*. Indian Institute of Science, Bangalore, India, August 30, 1999 (163–167).

Narayanasamy, P., & Gnanakumar Daniel, A. (1989). Lignite flyash: A non-polluting substance for tackling pest problems. In K. V. Devaraj (Ed.), *Progress in Pollution Research. Proceedings of Natinal Young Scientists' Seminar on Environmental Pollution*, University of Agricultural Sciences, Bangalore, India (201–206).

Raghupathy, B. (1988). *Effect of lignite flyash as a source of silica and phosphorus on rice, maize and sugarcane in lateritic soil*. Ph.D. thesis, Faculty of Agriculture, Annamalai University, Annamalai Nagar, Tamil Nadu, India.

Selvanarayanan, V., Ganesan, K., & Narayanasamy, P. (1997). Insecticidal activity of lignite flyash against the larvae of *Spodoptera litura* F. and *Crocidolomia binotalis* Z. In P. Narayanasamy (Ed.), *Flyash in Agriculture. Proceedings of National Seminar Seminar on Flyash*, Annamalai University, Annamalainagar, Tamil Nadu, India (8–11).

Selvi, K. (2008). Bio-management of termite, *Hypotermes obscuriceps* Wasmann (Isoptera: Termitidae). M.Sc. (Ag.) thesis, Annamalai University, Annamlainagar, Tamil Nadu, India.

Vijayakumar, N., & Narayanasamy, P. (2001). Safety of lignite flyash to non target organisms in rice ecosystem. In P. Narayanasamy (Ed.), *Agriclutural Application of Flyash. Proceedings of II National Seminar on Use of Flyash in Agriculture,* Annamalai University, Annamalainagar, Tamil Nadu, India (65–70).

Synthesis, Quality Assay and Assessment of Fly Ash-Based Chemical Pesticides for Efficacy against Pests of Crops, Stored Commodities and in Urban Areas

R. Ayyasamy, S. Sithanantham and P. Narayanasamy

Abstract Four fly ash-based insecticides, viz. Fly ash + Endosulfan 4% Dust, Fly ash + Chlorpyriphos 1.5% WDP, Fly ash + Imidacloprid 2.15% Tablet and Fly ash + Fipronil 0.05% Tablet, were synthesized. Fractions of lignite fly ash particles obtained through the ball mill grinder were characterized and fractions which were nearer to nanoparticle (less than 50 μm) were subjected to tests in various experiments. Maximum mean per cent mortality was found in Fly ash + Endosulfan 4% Dust 50% conc. against major pests of rice such as leaf folder, skipper, grasshopper, green horned caterpillar, green leafhopper and brown planthopper followed by Fly ash + Chlorpyriphos 1.5% WDP @ 0.25% which recorded mortality of leaf folder, rice skipper, grasshopper, green horned caterpillar, green leafhopper and brown planthopper. Maximum mortality was found in Fly ash + Endosulfan 4% Dust 50% conc. against major pests of cotton such as spotted bollworm, red cotton bug, aphid, mealy bug and whitefly followed by Fly ash + Chlorpyriphos 1.5% WDP @ 0.25% which recorded higher mortality of spotted bollworm, red cotton bug, aphid, mealy bug and whitefly. In brinjal, Fly ash + Endosulfan 4% Dust 50% conc. controlled major pests such as Epilachna beetle, aphid, mealy bug and lace wing bug followed by Fly ash + Chlorpyriphos 1.5% WDP @ 0.25%. Fly ash + Endosulfan 4% Dust @ 50% conc. gave better control of storage pests such as *Tribolium castaneum* (Herbst) and *Callosobruchus chinensis* L. followed by Fly ash + Chlorpyriphos 1.5% WDP @ 0.25%. In the choice test, cockroaches did not accept the test product FA + Imidacloprid, whereas they accepted the product FA + Fipronil to some extent and maximum number of cockroaches got attracted to groundnut candy (>3) in both the experiments and biscuits 2.62 in experiment one and 3.13 in experiment two compared to fipronil (1.5). Overall the preference tests had showed that the out of the two new fly-ash

R. Ayyasamy (✉) · P. Narayanasamy
Faculty of Agriculture, Department of Entomology, Annamalai University, Annamalainagar, Tamil Nadu 608 002, India
e-mail: ayyasamy.regu@gmail.com

P. Narayanasamy
e-mail: drpnsamy@yahoo.com

S. Sithanantham
Sun Agro Bio System Private Ltd., Porur, Chennai, India
e-mail: sabrcchennai@yahoo.co.in

S. K. Ghosh and V. Kumar (eds.), *Circular Economy and Fly Ash Management*,
https://doi.org/10.1007/978-981-15-0014-5_6

products, FA + Fipronil was found to be potential; however, it might be further enhanced through suitable refinement in the formulation.

Keywords Pesticide formulations · Pesticide carriers · Fly ash · Bio-efficacy

1 Introduction

Fly ash is a coal combustion product and waste generated from the thermal power stations with particle size range up to 100 μm (Davison et al. 1974). Though the utilization of fly ash in India has increased may folds up to 51 million tons, still sound management and disposal of fly ash have to be addressed. Scientists have paved way for better management practices/appropriate technologies towards safe utilization of the fly ash. Neyveli Lignite Corporation India Ltd. situated at Neyveli 11.5432°N, 79.4760°E Cuddalore District, Tamil Nadu State, India, generates approximately 38,485 metric tons of lignite fly ash. As the production site is close by our location, lignite fly ash generated from this thermal platformed the base material in our study.

Thus, the current study justifies the safe utilization of this solid waste material, fly ash as a carrier in formulating chemical pesticides in an economical manner to mitigate the environmental crisis generated by fly ash and cost involved in the inert carriers used in the formulations of insecticides.

2 Scenario of Chemical Pesticides in India

A pesticide is an agent that discourages and manages the damages caused by pest or kills. The noted author, Rachel Carlson in her book "Silent Spring" made popular the risks posed for by DDT (FAO 2005). Pesticides may be grouped in several ways such as based on target pests they destroy, for instance, insecticides, herbicides, rodenticides and others. They may be grouped according to chemical class they belong to, for example, organochlorines, organophosphorous, carbamates, pyrethroids, nitrophenols, neonicotinods. Details of consumption of pesticides to mitigate the pests in various parts of India are presented hereunder (Table 1).

Considering past experience of the Department of Entomology, Annamalai University on utilization of fly ash in pest control, fly ash unit, Department of Science and Technology, Govt. of India came forward to sanction a project, "Use of fly ash as a carrier in insecticide formulations" from which a set of four chemical pesticides were synthesized, efficacy assessed against pests of rice, cotton, vegetables and results are furnished.

Table 1 Chemical pesticides statewise consumption in India (Indira Devi et al. 2017)

S. No.	State/UT	Consumption (Tonnes)	Per ha consumption (kg/ha of gross cropped area)	CAGR (Per cent 2000–01 to 2012–13)
1	Andaman and Nicobar Islands	7	0.467	17.85
2	Andhra Pradesh	6500	0.581	4.26
3	Assam	183	0.065	−6.08
4	Arunachal Pradesh	17	0.080	−5.44
5	Bihar	687	0.131	−2.4
6	Chandigarh			
7	Chhattisgarh	675	0/144	3.29
8	Goa	9	0.069	5.42
9	Gujarat	1210	0.117	−6.24
10	Haryana	4050	1.151	−2.07
11	Himachal Pradesh	320	0.594	−0.26
12	Jammu and Kashmir	1711	2.337	100.22
13	Jharkhand	151	0.139	6.89
14	Karnataka	1225	0.116	−4.3
15	Kerala	856	0.413	−0.66
16	Madhya Pradesh	659	0.044	2.18
17	Maharashtra	6617	0.380	6.67
18	Manipur	30	0.086	5.25
19	Meghalaya	9	0.032	3.05
20	Mizoram	4	0.031	−7.53
21	Nagaland	16	0.044	8.82
22	Odisha	601	0.128	−2.08
23	Punjab	5725	1.377	−2.23
24	Rajasthan	1250	0.068	−1.78
25	Sikkim	3	0.039	1.13
26	Tamil Nadu	1919	0.387	1.37
27	Tripura	266	1.039	9.21
28	Uttar Pradesh	9035	0.545	2.95
29	Uttarakhand	220	0.304	7.29
30	West Bengal	3390	0.679	0.91
	Total	45,619	0.2912	–

3 Review of Research on Fly Ash

3.1 As a Pesticide

Major pests of rice and cotton crops were found susceptible to fly-ash dusting at 40 kg ha^{-1}. Fly-ash depressed the larvae of crop pests by imparting damage to their feeding organs and digestive system and induced plant resistance against brown planthopper and yellow stem borer in rice (Narayanasamy 2002). Application of 5% fly-ash to soil enhanced the growth and vigour of tomato plants which reflected in the reduced damage caused by root-knot nematode as the galling on the roots were much reduced (Ahmad and Mashkoor Alam 1997). Plots treated with 30% fly-ash reduced the population and reproductive ability of the root-knot nematode in tomato ecosystem (Khan et al. 1997).

3.2 As a Carrier in Pesticide Formulations

For the first time, Vijayakumar and Narayanasamy (1995) synthesized four insecticides, viz. BHC 10% Dust, BHC 50% WDP, Malathion 25% WDP and Carbofuran 3% WDG involving fly ash as a carrier. Results exhibited that all the fly ash-based pesticides excelled their respective commercial chemicals in controlling leaf folder, yellow hairy caterpillar and sucking pests, namely brown planthopper, green leafhopper and earhead bug. Fly-ash-based herbal pesticides developed were found effective against pests of rice, brinjal and bhendi (Arputhasankari and Narayanasamy 2007).

Further of fly-ash was found to be a termiticide, and topical application of the fly-ash on termites was found superior over other modes of application such as fly-ash paste, fly-ash-dusted wood and fly-ash + moist soil mixture against worker castes of termites (Selvi 2008).

4 A Look at Carriers in Pesticide Formulations

A pesticide formulation comprises both of active technical grade compound and inert ingredients. Active ingredients are meant to kill or drive away the pests. Other ingredients used are to stabilize the product and extend shelf life, to enhance the wetting and spreading quality on the surface applied, prevent caking or foaming, ease of applications, make ingredients compatible and aid in drift control (NPIC 2001).

A mineral carrier provides a safe, effective and economical mean for distributing the pesticides in all our agricultural practices. The minerals most frequently encountered include montmorillonite, attapulgite and diatomite. The crystal structures of adsorptive carriers contribute to their performance. Mineral carriers come in various bulk densities and particle size distribution.

Salient features of some of the important minerals serving as carrier in pesticide formulations are presented hereunder (Table 2).

Table 2 Summary features of important carrier materials used in pesticide formulations (Yusoff et al. 2016)

Material	Used with active ingredient	Class	Remarks
Silicon dioxide nanoparticles	a-Pinene and Linalool	Antifeedant	Enhances the antifeedant properties of a-pinene and Linalool. Stable up to 6 months storage
Ultrafine fibre cellulose	Avermectin	Pesticide	Controls the release of avermectin in 30% of ethanol solution
Cyanobacteria	Avermectin	Pesticide	Prolongs the release of avermectin on mixture of ethanol/water(4:1 v/v) protects avermectin against UV radiation
b-cyclodextrin	Capsaicin	Pesticide	Enhances the solubility of capsaicin by fivefold. Improves the capsaicin stability against light and heat
Chitosan copolymer with poly(lactide) and 1,2-dipalmitoyl-sn-glycero-3-phosphoethanolamine	Chlorpyrifos	Insecticide	Effective sustain release of chlorpyrifos
Straw ash-based biochar and biosilica	Chlorpyrifos	Insecticide	Controls the loss of chlorpyrifos due to washing, volatilization and leaching to soil
Chitosan-coated beeswax solid lipid	Deltamethrin	Insecticide	Good control release properties of deltamethrin and protection against UV radiation
Starch–silver nanoparticle	Dichlorvos and chlorpyrifos	Insecticide	Enhances the encapsulation efficiency of the insecticides up to 95–98%. Slow-release properties on target pests up even after 21 days of application

(continued)

Table 2 (continued)

Material	Used with active ingredient	Class	Remarks
Silica nanocapsules from tetraethoxysilane	Fipronil	Insecticide	High encapsulation efficiency of fipronil (70% encapsulation efficiency)
Alginate nanoparticles	Imidacloprid	Insecticide	Encapsulate imidacloprid for about 98.66%. Effectively control the population of leafhopper on Okra crop up to 15 days. Less cytotoxicity as compared to commercial pesticide
Polylactic acid	Lambda-cyhalothrin	Pesticide	Microcapsule delivery system have better water dispersion and give sustained release behaviour
Lignin–alginate and lignosulphonate	Pyrethrins	Insecticide	Enhances the photo stabilisation and reduces the volatilization of pyrethrins
Poly(e-caprolactone)	Simazine	Pesticide	Slow release of simazine in water. Reduces the toxicity effect of simazine
Nanosized calcium carbonate	Validamycin	Pesticide	Sustains the release of validamycin. Good germicidal efficacy against Rhizoctonia solani for about 2 weeks

4.1 Clay Minerals

Minerals with less than 2 μm are described as clay minerals. The clay minerals have specific surface and contain inorganic cation, higher stability and a variety of structural properties because of their many layered structure. Clay minerals are superior adsorbents for ionic and polar compounds and very low toxicity effects (Choy et al. 2007)

Clay minerals such as bentonite, kaolinite and sepiolite used as carrier in pesticide synthesis can enhance the stability and shelf life of the pesticides.

4.2 Siliceous Materials

Siliceous materials, such as glass, zeolite, have been proposed to modulate pesticide products (Chen et al. 2011), and their availability in ample quantities makes them appropriate materials for pesticide production.

In recent times, silicon dioxide nanoparticles have been used to formulate botanical active ingredients such as linalool and a-pinene (Rani et al. 2014). A formulation which contains linalool and a-pinene with the nanoparticles had enhanced the antifeedant and keeping quality of the chemicals.

4.3 Cyclodextrin

Cyclodextrin is a "Macrocyclic oligo sugar" product of bacterial degradation of starch which comprises of six, seven or eight glucose units (alpha, beta and gamma cyclodextrins) which are the products of bacterial degradation of starch. In recent times, cyclodextrin has attracted the attention of chemists for plausible role in pesticide formulations.

4.4 Chitosan

Chitosan, a product of the chemical deacetylation of chitin and most abundant natural polymers next to cellulose, has strong biological properties such as insecticidal and has been predominantly used in agriculture. Chitosan can be easily modified as to result in formulations of various chitosan derivatives in order to widen the spectrum of usage of chitosan. Photodegradation of deltamethrin can be lowered if coated with chitosan-coated beeswax solid lipid. Carboxymethyl chitosan was used as carrier for preparing a botanical pesticide, azadirachtin (Feng and Peng 2012). Novel nanoparticles based on octahydrogenated retinoic acid conjugated to glycol chitosan developed by Lu et al. (2013) have confirmed the potential sustained release profile of azadirachtin.

4.5 Others

Natural and synthetic polymeric matrices, such as cellulose, lignin and sodium alginate, have received great attention in formulations of the pesticides. The potential of alginate as a carrier for imidacloprid was studied well. Lignin and lignosulphonate-based formulations protect insecticidal pyrethrins against photodegradation and volatilization.

Organic and inorganic materials received interest in the pesticide formulation sector as they exhibit great potential in pesticide delivery systems. Use of cyanobacteria an environmental waste has been found as a potential carrier for the pesticide avermectin. Biochar, a product, is a thermal decomposition process of biomass under oxygen-limited conditions. A novel pesticide formulation for chlorpyrifos based upon straw ash-based biochar and biosilica (BCS) was developed.

Nanosized calcium carbonate has been used as a carrier for the formulating validamycin.

5 Materials and Methods

5.1 Fly Ash Collection

Lignite fly ash generated from thermal power stations of the neighbouring Neyveli Lignite Corporation Ltd., Neyveli was procured and subjected in our study.

5.2 Laboratory Analyses of Fly ash

Certain physical and chemical parameters of fly-ash were analysed in the laboratory such as bulk density, moisture content, water holding capacity, cation exchange capacity and E.C using standard laboratory protocols. pH was determined in soil to water ratio of 1:2.5, organic carbon by rapid dichromate oxidation technique, determination of Mn, Cu and Zn by HNO_3 and $HClO_4$ by digestion method, sulphate estimation by turbidimetric method, estimation of Ca and Mg by EDTA titrimetric method, determination of N by Kjeldahl method, exchangeable potassium by centrifugation and decantation procedure determination of available phosphorus by Bray's method and heavy metal detection by DTPA extractable method (Maiti 2003).

5.3 Characterization of Lignite Fly Ash Dust of Various Fractions

Fractions of lignite fly ash particles obtained through the ball mill grinder were characterized for morphological features. All the fractions entertained were irregular in shape. As per standardization made by Prof. P. Narayanasamy at the Department of Entomology, Annamalai University, a fraction which is nearer to nanoparticle (less than 50 μm) was subjected to tests in the project.

5.4 Production of 100% Fly Ash Dust

Two hundred and fifty kilograms of fly ash-based pesticides were synthesized with different micron sizes that were obtained for our experiments.

Size	Quantity (kg)
210 μm	90.00
125 μm	55.00
105 μm	55.00
Undesignated–'X'	50.00
Total	250.00

The fine neutralized fly ash was packaged in autoclavable propylene polybags and sterilized in an autoclave for 4 h continuously at 121 °C or 15 psi.

5.5 Blending

Pesticides were blended with fine neutralized and sterilized fly ash. Four fly ash-based, Fly ash + Endosulfan 4% Dust, Fly ash + Chlorpyriphos 1.5% WDP, Fly ash + Imidacloprid 2.15% Tablet and Fly ash + Fipronil 0.05% Tablet were synthesized.

5.6 Synthesis of Fly Ash-Based Tablet Insecticides

Two fly-ash-based insecticides, namely FA + Imidacloprid 2.15%, FA + Fipronil 0.05%, were formulated. Their compositions and their proportion in each are given below:

Imidacloprid 2.15% tablet		
Constituents	Quantity/100 g	Per cent
Fly ash	45.0 g	45.00
Imidacloprid (a.i)	2.5 g	02.50
Boric acid	10.0 g	10.00
Naphthalene sulphonate	10.0 g	10.00
Starch	25.0 g	25.00
Solvent	2.5 ml	2.50

(continued)

(continued)

Gum	5.0 ml	5.00
Fipronil 0.05% tablet		
Constituents	Quantity/500 g	Per cent
Fly ash	225.0 g	45.00
Fipronil (a.i)	2.5 g	00.50
Boric acid	65.0 g	13.00
Starch	125.0 g	25.00
Gum	25.0 g	05.00
Naphthalene sulphonate	50.0 g	10.00
Solvent	7.5 ml	01.50
Gum	5.0 ml	01.00

The ingredients were blended with pestle and mortar and poured into tablet moulds; air-dried for 48 h and then sealed in aluminium foil.

5.7 *Bioassay Methodology Adopted*

Bio-efficacy tests conducted in the laboratory were focused to assess their acceptability by and the extent of mortality caused to the test insect. To simulate the natural settings, the candidate products were individually compared with alternative food sources (preference test). To verify if they would be accepted in the absence of alternatives as feed, they were also kept alone (no choice test).

Adults of the domestic cockroach (*Periplaneta americana* L.) were collected from natural sources (municipal drainage pipes in urban centres) and subjected to one-day stabilization in the laboratory by keeping along with broken pieces of biscuits as feed and moisture provided from water-soaked balls made of newspaper. For each replicate and treatment, 20 adults were assigned.

5.7.1 Preference Test of the Fly ash-Based Pesticides

Two test pesticides, Imidacloprid and Fipronil (A, B), were each kept along with alternative insect feed sources (C-Biscuits, D-Groundnut candy) in individual containers (plastic troughs of 30 cm dia., 10 cm depth). The treatments compared were as follows with 10 replicates in each.

Acceptance of the test products

At 1 week after the starting of the test, as an index of acceptability for feeding, the extent of consumption of the product tablets (A/B) was visually rated on 0–5 scale as follows: in comparison with the other feeds (C/D) kept in the same container.

Rating scale

0 = No feeding; 1 = Light feeding (below 10%); 2 = Moderate feeding (11–30%); 3 = Substantial feeding (31–50%); 4 = Considerable feeding (51–75%); 5 = Heavy feeding (Above 75%).

Evaluation of efficacy

As a combined index of the extent of feeding and relative toxicity of the product, the mortality was recorded (among the 20 adults) in each container at 1st, 3rd, 5th and 7th days after the start of the experiment. The per cent adult mortality was also worked out for the products.

5.7.2 No Choice Test

In each trough, the test product alone (A or B) was kept. There was no alternate feed kept, except moist filter paper for keeping up humidity. For each set, ten replicates were kept, along with untreated control. The data on the mortality of the test insect was recorded as in the earlier study.

6 Results and Discussion

6.1 Evaluation of Lignite Fly ash 100% Dusts of Various Particle Sizes Against Larvae of Spodoptera litura *Fabricius (Table 3)*

Undesignated fractions of fly ash dust exhibited the highest per cent mortality of 63.44 followed by 210 μm (58.91%), 105 μm (50.77%) and 125 μm (43.09). Larvae treated

Table 3 Effect of Lignite Fly ash 100% Dust of Various Particle Sizes Against Larvae of *Spodoptera litura* F.

S. No.	Treatments	Mean per cent mortality of larvae*
1	210 μm	58.91b
2	125 μm	43.09d
3	105 μm	50.77c
4	Undesignated	63.44a
5	Untreated control	0.0e
SE		1.84
Cd ($p = 0.05$)		3.96

*Arc sine transformed values—mean of four replications
Values followed by different alphabets vary significantly by DMRT

Plate 1 Effects of application lignite fly ash 100% dust with particle size of undesignated on *Spodoptera litura* larvae by poisoned food technique

with undesignated fractions of fly ash exhibited larval pupal mosaic, dead pre-pupa and malformed pupa (Plate 1).

6.2 Synthesis and Evaluation of Fly ash-Based Insecticide Formulations in Laboratory

After witnessing efficacy of the fly ash dust 100% against various insect pests, the fly ash was used as a carrier in synthesizing two insecticides formulations, viz. FA + Chlorpyriphos 1.5%, FA + Endosulphan 4% Dust. Two more new formulations were evaluated for their efficiency against various pests of the target crops, like rice, brinjal and cotton. It was evident that both the fly ash insecticides were superior to the commercial insecticides.

It is concluded that both the fly ash insecticides, viz. FA + Chlorpyriphos 1.5% WDP and FA + Endosulphan 4% Dust, showed their might in checking various pests infesting crops like rice, cotton and brinjal.

It was found that both the new formulations showed themselves slightly superior to the chemical insecticides against certain key pests of grains under storage condition.

6.3 Efficacy of Fly Ash-Based Insecticides Against Major Pests of Rice

Mean per cent mortality was found maximum in Fly ash + Endosulfan 4% Dust 50% conc. against major pests of rice such as leaf folder (67.34), skipper (74.80), grasshopper (59.69 green horned caterpillar (59.81), green leaf hopper (63.33) and brown planthopper (65.03) followed by Fly ash + Chlorpyriphos 1.5% WDP @ 0.25% which recorded mean per cent mortality of leaf folder (65.36), rice skipper (71.71), grasshopper (57.62) green horned caterpillar (57.81), green leafhopper (62.00) and brown planthopper (62.78). Fly ash-based formulations of insecticides were found superior when compared to conventional insecticides (Table 4).

6.4 Efficacy of Fly Ash-Based Insecticides Against Major Pests of Cotton

Mean per cent mortality was found maximum in Fly ash + Endosulfan 4% Dust 50% conc. against major pests of cotton such as spotted bollworm (77.33), red cotton bug (67.23), aphid (67.58), mealy bug(78.99) and whitefly (67.23) followed by Fly ash + Chlorpyriphos 1.5% WDP @ 0.25% which recorded mean per cent mortality of spotted bollworm (74.71), red cotton bug (65.33), aphid (65.83), mealy bug (77.61) and whitefly (65.33). Fly ash-based formulations of insecticides were found superior when compared to conventional insecticides (Table 5).

6.5 Efficacy of Fly Ash-Based Insecticides Against Major Pests of Brinjal

Mean per cent mortality was found maximum in Fly ash + Endosulfan 4% Dust 50% conc. against major pests of brinjal such as Epilachna beetle (74.11), aphid (71.23), mealy bug (62.10) and lace wing bug (62.79) followed by Fly ash + Chlorpyriphos 1.5% WDP @ 0.25% which recorded mean per cent mortality of Epilachna beetle (71.28), aphid (68.74), mealy bug (59.91) and lace wing bug (60.56). Fly ash-based formulations of insecticides were found superior when compared to conventional insecticides (Table 6).

Table 4 Bio-efficacy of fly ash-based chemical pesticides against major pests of rice under laboratory conditions

Treatments	Mean per cent mortality of					
	Rice leaf folder	Rice skipper	Rice grasshopper	Rice green horned caterpillar	Rice green leaf hopper	Rice brown plant hopper
FA + Chlorpyriphos 1.5% WDP @ 0.1% Conc.	61.49	68.27	54.11	53.11)	59.66	58.84
FA + Chlorpyriphos 1.5% WDP @ 0.25% Conc.	65.36	71.71	57.62	57.81	62.00	62.78
Commercial Chlorpyriphos 1.5% WDP @ 0.1% Conc.	52.68	60.09	46.03	45.33	51.44	51.73
Commercial Chlorpyriphos 1.5% WDP @ 0.25% Conc.	56.75	64.22	49.82	49.99	56.07	55.00
FA + Endosulfan 4% D @ 30%	63.11	70.57	55.67	55.14	60.44	60.64
FA + Endosulfan 4% D @ 50%	67.34	74.80	59.69	59.81	63.33	65.03
Commercial Endosulfan 4% D @ 30%	54.73	61.31	47.67	48.15	54.26	53.11
Commercial Endosulfan 4% D @ 50%	59.73	66.24	51.33	51.76	57.62	56.59
FA 100% Dust	50.00	60.00	45.00	43.33	50.00	48.66
Untreated control	0.00	0.00	0.00	0.00	0.00	0.00
CD ($p = 0.05$)	0.569	0.597	0.012	0.003	0.012	0.016
S.E.D.	0.283	0.297	0.006	0.001	0.006	0.008

Mean of three replications

Table 5 Bio-efficacy of fly ash-based chemical pesticides against major pests of cotton under laboratory conditions

Treatments	Mean per cent mortality of				
	Spotted bollworm	Red cotton bug	Aphid	Mealy bug	Whitefly
FA + Chlorpyriphos 1.5% WDP @ 0.1% Conc.	71.83	60.72	61.11	74.11	60.72
FA + Chlorpyriphos 1.5% WDP@ 0.25% Conc.	74.71	65.33	65.83	77.61	65.33
Commercial Chlorpyriphos 1.5% WDP @ 0.1% Conc.	63.21	54.76	54.16	65.56	54.76
Commercial Chlorpyriphos 1.5% WDP @ 0.25% Conc.	67.09	56.78	58.50	70.09	56.78
FA + Endosulfan 4% D @ 30%	72.81	63.36	63.21	76.73	63.36
FA + Endosulfan 4% D @ 50%	77.33	67.23	67.58	78.99	67.23
Commercial Endosulfan 4% D @ 30%	67.09	54.76	56.22	67.21	54.76
Commercial Endosulfan 4% D @ 50%	69.66	58.23	60.00	72.38	58.23
FA 100% Dust	60.33	45.10	51.66	60.50	45.10
Untreated control	0.00	0.00	0.00	0.00	0.00
CD ($p = 0.05$)	0.022	0.009	0.451	0.136	0.023
S.E.D.	0.011	0.004	0.224	0.068	0.011

Mean of three replications

6.6 *Efficacy of Fly Ash-Based Insecticides Against Storage Pests*

Mean per cent mortality was found maximum in Fly ash + Endosulfan 4% Dust 50% conc. against certain storage pests such as *Tribolium castaneum* (Herbst) (46.67) and *Callosobruchus chinensis* L. (60.00) followed by Fly ash + Chlorpyriphos 1.5% WDP @ 0.25% which recorded mean per cent mortality of *T. castaneum* (39.33) and

Table 6 Bio-efficacy of fly ash-based chemical pesticides against major pests of brinjal under laboratory conditions

Treatments	Mean per cent mortality of			
	Epilachna beetle	Aphid	Mealy bug	Lace wing bug
FA + Chlorpyriphos 1.5% WDP @ 0.1% Conc.	67.34	63.67	56.46	56.68
FA + Chlorpyriphos 1.5% WDP @ 0.25% Conc.	71.28	68.74	59.91	60.56
Commercial Chlorpyriphos 1.5% WDP @ 0.1% Conc.	60.07	54.29	47.33	51.44
Commercial Chlorpyriphos 1.5% WDP @ 0.25% Conc.	63.74	58.75	51.11	54.71
FA + Endosulfan 4% D @ 30%	69.23	65.34	59.02	58.59
FA + Endosulfan 4% D @ 50%	74.11	71.23	62.10	62.79
Commercial Endosulfan 4% D @ 30%	61.26	56.66	49.66	51.44
Commercial Endosulfan 4% D @ 50%	65.35	61.42	55.34	55.70
FA 100% Dust	60.00	50.00	42.90	50.00
Untreated control	0.00	0.00	0.00	0.00
CD ($p = 0.05$)	4.495	0.008	0.006	0.011
S.E.D.	2.236	0.004	0.003	0.005

C. chinensis (49.83). Fly ash-based formulations of insecticides were found superior when compared to conventional insecticides (Table 7).

The overall order of toxicity was FA + Endosulfan 4% D @ 50% > FA + Chlorpyriphos 1.5% WDP @ 0.25% conc. > FA + Endosulfan 4% D @ 30% > FA + Chlorpyriphos 1.5% WDP @ 0.1% conc. > Commercial Endosulfan 4% D @ 50% > Commercial Chlorpyriphos 1.5% WDP @ 0.25% conc. > Commercial Endosulfan 4% D @ 30% > Commercial Chlorpyriphos 1.5% WDP @ 0.1% conc. > FA 100% Dust.

6.7 *Choice for Feeding Acceptance Study*

The preference for feeding study (choice test) showed that in the presence of alternate food, the test insects did not accept the test product FA + Imidacloprid (A), whereas they accepted the product FA + Fipronil, B to some extent (Table 8).

FA + Imidacloprid, Imidacloprid in particular seemed to be highly repulsive, apparently due to the offensive smell of the insecticide molecule, whereas the product FA + Fipronil, Fipronil appears somewhat acceptable, but may perhaps be improved further for acceptability by some minimal refinement. Accordingly, the two presently

Table 7 Bio-efficacy of fly ash-based chemical pesticides against storage pests

Treatments	Mean per cent mortality	
	T. castaneum	*C. chinensis*
FA + Chlorpyriphos 1.5% WDP @ 0.1% Conc.	35.89	53.17
FA + Chlorpyriphos 1.5% WDP@ 0.25% Conc.	34.39	46.17
Commercial Chlorpyriphos 1.5% WDP @ 0.1% Conc.	39.33	49.83
Commercial Chlorpyriphos 1.5% WDP @ 0.25% Conc.	36.11	50.33
FA + Endosulfan 4% D @ 30%	41.83	51.50
FA + Endosulfan 4% D @ 50%	42.28	56.67
Commercial Endosulfan 4% D @ 30%	21.83	38.50
Commercial Endosulfan 4% D @ 50%	46.67	60.50
FA 100% Dust	26.16	42.67
Untreated control	0.00	0.00
CD ($p = 0.05$)	0.599	0.692
S.E.D.	0.298	0.344

Mean of three replications

Table 8 Preference test—the extent of feeding by cockroaches on different feed materials at different time intervals

Treatments	No. of cockroaches attracted to feed on				
	Day 1	Day 3	Day 5	Day 7	Mean
Experiment 1					
1. Imidacloprid	0.0a	0.0a	0.0a	0.0a	0.00
2. Biscuits	1.5b	2.5b	3.0b	3.5b	2.62
3. Groundnut candy	2.5c	3.0b	3.5b	3.5b	3.13
Experiment 2					
1. Fipronil	0.5a	1.5a	2..0a	2.0a	1.5
2. Biscuits	2.0b	3.0b	3.5b	4.0b	3.13
3. Groundnut candy	2.5b	3.5b	4.0b	4.0b	3.5
LSD ($p = 0.05$)	0.45	0.84	0.95	1.18	–

formulated products might not be ready to be immediately commercialized and might require further R&D to reformulate them suitably to suppress this repulsiveness, possibly by masking with some additional attractant ingredient so as to enable the target insects (cockroach) to be attracted to feed on the new fly ash-based product. While such reformulation/refinement is apparently very much necessary for the Product FA + Imidacloprid, the product FA + Fipronil, Fipronil appeared to be somewhat satisfactory but might also benefit through enhancing the attractiveness/acceptability for feeding. Mean number of cockroaches attracted was higher on groundnut candy (>3 in both the experiments) and biscuits (2.62 in experiment one and 3.13 in experiment two) compared to fipronil (1.5) and imidacloprid did not attracted the test insect.

6.8 Choice Test: Cumulative Mortality of Cockroaches Due To Fly ash-Based Insecticides

The mortality observed with the test insects with Imidacloprid and Fipronil (Table 9) exhibited a significant difference between the two products FA + Imidacloprid and FA + Fipronil as the mean mortality of Imidacloprid and Fipronil recorded was 5 and 20%, respectively.

The trend in the result was on the expected line since the toxicants incorporated in the test product FA + Imidacloprid could not even be ingested due to its apparent repulsiveness, and hence, there was insignificant mortality. On the other hand, the product FA + Fipronil was consumed to a limited extent, culminating in modest levels of mortality over the experimental period. Of course, the presence of the alternative foods (biscuits, groundnut candy) provided a near-realistic situation as would prevail in practice, since the domestic insects might not often be without access to alternative feed, which might serve as a source of weakening the impact of these fly-ash products. Nevertheless, the potential of product FA + Fipronil as an useful fly-ash-based formulation also needs to be improved, so as to stand to the requirements for commercialization.

Overall, the preference tests had showed that the out of the two new fly-ash products, product FA + Imidacloprid, Imidacloprid needs to be necessarily improved by reformulation to make it more attractive or less repulsive, whichever maybe the associated reason for the insects not feeding on the product. For the product FA + Fipronil, its potential might be further enhanced through suitable refinement in the formulation.

Table 9 Choice test—mortality of cockroaches for certain fly ash-based insecticides

Treatment	Per cent mortality on				
	Day 1	Day 3	Day 5	Day 7	Mean
1. Imidacloprid	5.0	5.0	5.0	5.0	5.00
2. Fipronil	10.0	20.0	25.0	25.0	20.00
3. Untreated Control	0.0	1.0	2.0	3.5	1.62
F test	NS	6.0	7.5	8.0	–

Table 10 No choice test—mortality of cockroaches due to certain fly ash-based insecticides

Treatment	Per cent mortality on				
	Day 1	Day 3	Day 5	Day 7	Mean
1. Imidacloprid	3.0	5.0	12.0	15.5	8.87
2. Fipronil	25.0	35.0	40.0	40.0	35.00
3. Biscuits	3.0	5.0	7.0	12.5	6.87
4. Groundnut candy	2.5	4.5	7.5	10.0	6.12
LSD ($p = 0.05$)	12.5	13.5	15.0	15.5	14.12

6.9 No Choice Test—Maximum Scope for Mortality of Cockroaches Due to Fly ash Insecticides

The results from the "No Choice test" showed significant differences in the mortality due to the two products, when confined with the individual test products alone (Table 10).

At the first interval (day 1) alone, the differences were not so strong, among all the treatments except fipronil. Subsequently, FA + Fipronil were able to show significant improvements in cumulative mortality.

These results have shown that the FA + imidacloprid were not apparently attractive enough to induce the target insect to feed upon, even in the absence of alternative food. In the case of product FA + Fipronil, there is scope, but further refinement to minimize deterrence/enhance attractiveness warranted is useful to make it commercially promising. The mean mortality of cockroaches was highest in fipronil (35%), and in other treatments, it ranged between 6.12 and 8.87%. The results also reconfirmed the need to improve the formulation protocol for these two candidate products, so as to include more strong feeding attractants in the recipe. Apparently, the proportion/source of the presentation included jigger as an attractant should be further improved in future attempts in the R&D.

6.10 Cost Economics of the Fly ash-Based Insecticides

S. No.	Name	Fly ash pesticides (US$)	Commercial pesticides (US$/kg)
1	Endosulphan 4% Dust	0.0260	0.0071
2	Chlorpyriphos 1.5% WDP	0.370	1.711
3	Fly ash 100% Dust	0.007	–

From the above, it is evident that the lignite fly ash had served as an excellent carrier in the formulation of two insecticides, Chlorpyriphos 1.5% WDP and Endosulphan 4% Dust. Aside from the efficacy of the said new insecticides as seen from the findings, lignite fly ash cuts down cost of the newly synthesized insecticides by 5–10 times as compared to that of the commercial insecticides. This is due to the fact that the fly ash is available enormously and is being sold out at a paltry cost of US$7/ton especially for industrial purposes like manufacture of cement, brick, concrete blocks. It is thus worthwhile to initiate efforts to usher in fly ash-based insecticides for cost effective plant protection in agriculture, which alone, it is hoped, will enable the farmers for gainful production of their produces.

7 Conclusion and Recommendations

From the findings obtained as above, it is evident that the two fly ash-based insecticide formulations, namely FA + Chlorpyriphos 1.5% WDP and FA + Endosulfan 4% Dust, are striking compounds recommended for tackling various pest problems in crops like rice, cotton and brinjal and other crops.

New insecticides formulated in the project were superior to the respective chemical insecticide formulations available in the market.

Therefore, the fly ash-based insecticides have potentials to manage household pests such as cockroaches and similar pest problems at home.

Acknowledgements We are grateful to Dr. Vimal Kumar, Former Head, Fly ash Unit, Department of Science and Technology (DST), Government of India for granting the project with substantial funds, and we owe a lot to DST for the support. We are indebted to authorities of Annamalai University for necessary permission and provision of facilities for the conduct and completion of the project in time. We thank wholeheartedly Dr. B. V. David, Former President, M/s Sun Agro Biosystem Private Ltd., Porur, Chennai, Tamil Nadu, India, for ready acceptance to be a partner in the project and for conducting a part of the project, and we express our deep sense of personal thanks to Dr. S. Sithanantham, Director, Sun Agro Biosystem Private Ltd., Porur, Chennai, for the continued project study as Co-Investigator and excellent co-operation and assistance.

References

Ahmad, A., & Mashkoor Alam, M. (1997). Utilization of fly ash and *Paecilomyces lilacinus* for the control of *Meloidogyne incognita*. *International Journal of Nematology, 7,* 162–164.

Arputhasankari, S., & Narayanasamy, P. (2007). Bio-efficacy of flyash-based herbal pesticides against pests of rice and vegetables. *Current Science, 92,* 811–816.

Chen, J., Wang, W., Xu, Y., & Zhang, X. (2011). Slow-release formulation of a new biological pesticide, pyoluteorin with mesoporous silica. *Journal of Agricultural and Food Chemistry, 59*(1), 307–311.

Choy, J. H., Choi, S. J., Oh, J. M., & Park, T. (2007). Clay minerals and layered double hydroxides for novel biological applications. *Applied Clay Science, 36*(1–3), 122–132.

Davison, R. L., Natusch, D. F., Wallace, J. R., & Evans, C. A. (1974). Trace elements in fly ash. Dependence of concentration on particle size. *Environmental Science and Technology, 8* (13), 1107–1113.

FAO. (2005). *Codex Alimentarius*. www.codexalimentarius.net. Accessed September 2018.

Feng, B. H., & Peng, L. F. (2012). Synthesis and characterization of carboxymethyl chitosan carrying ricinoleic functions as an emulsifier for azadirachtin. *Carbohydrate Polymers, 88,* 576–582.

Indira Devi, P., Judy, T., & Raju, R. K. (2017). Pesticide consumption in India: A spatiotemporal analysis. *Agricultural Economics Research Review, 30*(1), 163–172.

Khan, M. R., Khan, M. W., & Singh, K. (1997). Management of root knot disease of tomato by the application of fly ash in soil. *Plant Pathology, 46*(1), 33–43.

Lu, W., Lu, M. L., Zhang, Q. P., Tian, Y. Q., Zhang, Z. X., & Xu, H. H. (2013). Octahydrogenated retinoic acid-conjugated glycol chitosan nanoparticles as a novel carrier of azadirachtin: Synthesis, characterization, and in vitro evaluation. *Journal of Polymer Science Part A: Polymer Chemistry, 51*(18), 3932–3940.

Maiti, S. K. (2003). *Handbook of methods in environmental studies Vol. 2: Air, noise, soil & overburden analysis*. Jaipur, India: ABD Publishers.

Narayanasamy, P. (2002). *Lignite fly ash as eco-friendly insecticide*. Available from: http://www.hinduonnet.com. Accessed August 2008.

NPIC. (2001). Inert or other ingredient. Topic Fact sheet. National Pesticide Information Centre, U.S.A.

Rani, P. U., Madhusudhanamurthy, J., & Sreedhar, B. (2014). Dynamic adsorption of α-pinene and linalool on silica nanoparticles for enhanced antifeedant activity against agricultural pests. *Journal of Pest Science, 87*(1), 191–200.

Selvi, K. (2008). *Bio-management of termite Hypotermes obscriceps (Wasman) (Isoptera: Macrotermitidae)*. M.Sc. (Ag.) thesis, Department of Entomology, Annamalai University, Annamalai Nagar, Tamil Nadu, India.

Vijayakumar, N., & Narayanasamy, P. (1995, December). Use of fly ash as a carrier in insecticide formulation. In *Fly Ash in Agriculture, Proceedings of National Seminar on Use of Lignite Fly Ash in Agriculture, Annamalai Nagar, India* (28–30).

Yusoff, S. N. M., Kamari, A., & Aljafree, N. F. A. (2016). A review of materials used as carrier agents in pesticide formulations. *International Journal of Environmental Science and Technology, 13*(12), 2977–2994.

Potential and Futuristics of Fly Ash Nanoparticle Technology in Pest Control in Agriculture and Synthesis of Chemical and Herbal Insecticides Formulations

P. Narayanasamy

Abstract Fly ash, one of the numerous substances that causes air, water and soil pollution, has been discovered as a pesticide against various pest forms that hamper agriculture and an active carrier in chemical pesticides (dust, wettable powder, granules and capsules) and herbal insecticides (dust) formulations. Insecticidal action of the fly ash dust was detected for the first time in the country on various kinds of insect pests inflicting crops like rice, vegetables, oilseeds, pulses, greens, fruit trees, besides pests of grains in storage yards too. Mode of action of fly ash dust on insect forms was characterised based on its physical and internal damage both on the insect bodies and crop plants. Interestingly, fly ash on application to rice soils had promoted resistance to the plant to thwart pests attack. Nano-fly ash of two fractions of size ranging down 50 μm was detected to be highly potential in killing the pests. Four chemical insecticides, viz. BHC 10% Dust, BHC 50% WDP, Malathion 25% WDP and Carbofuran 3% WDG, were synthesized with the two selected lignite fly ash fractions. All of them faired best in killing various pest's species in rice compared to the respective chemical insecticides. It is felt that use of fly ash as a carrier in the synthesis of chemical and herbal pesticides may replace with long run the conventional carriers like Calcite, Magnesite for eventual decline in the cost, and hence, will be a breakthrough in the pesticide industry. As an attempt to exploit particle, morphology and mineralogical contents of fly ash of two sizes ranging from among 10 to 50 μm were selected for their increased pesticidal action. SEM studies carried to have a close-up view of the individual particle of the lignite fly ash and coal fly ashes revealed that they were mostly of nanoparticle types. Secondly, morphometric features of the lignite fly ash nanoparticles were of spherical shape containing mostly of silica as silicon di oxide (SiO_2) and showed two forms, namely amorphous which is rounded and smooth and crystalline which is sharp and pointed. These differently sized particles of fly ash are best suited to adhere to the body skin of the insects having dense cover of structures like fine hairs, scales, spine-like processes, nodules, pustules, ventricles. Further, such fly ash particles when delivered in the field through dusters cling firmly to the plants and the bodies of insects and per cent deposition

P. Narayanasamy (✉)
Department of Entomology, Faculty of Agriculture, Annamalai University, Annamalainagar, Tamil Nadu 608 002, India
e-mail: drpnsamy@yahoo.com

S. K. Ghosh and V. Kumar (eds.), *Circular Economy and Fly Ash Management*,
https://doi.org/10.1007/978-981-15-0014-5_7

was high. Mineral contents of the fly ash revealed the presence of Silica, Alumina, Calcium, Ferric Oxide and traces of Zn, Pb, Zr, Sr, S, Th, Cu, Mn, etc. Among these, silica has been observed to strengthen the pesticide property of fly ash followed by Al, Ca, Fe and sulphur. From the above, it is inferred that such nano-fly ash particle technology has great scope in pest control in agriculture and allied arenas of farm folks through promoting organic agriculture tactics.

Keywords Fly ash nanoparticle · Exoskeleton · Adhesion · Adjuvant · Pesticide formulations

1 Introduction

Fly ash, one of the numerous substances that cause air, water and soil pollution so far, has been discovered as a pesticide and an adjuvant in pesticide formulations, viz. dust, wettable powder, granule, capsules, besides useful in various sectors in agricultural fields. This novel finding emanated through series of research projects spanning nearly 25 years funded by Department of Science and Technology, and Coal, Government of India, New Delhi, and Tamil Nadu State Council for Science and Technology, Government Tamil Nadu, undertaken under my guidance at Annamalai University.

2 Generation and Utilization Scenario of Fly ash in India

China, USA, Germany and India are the biggest producers of fly ash in the world. It is learnt that the current annual production of fly ash in India has exceeded 180 mill. tonnes emitting from 120 odd thermal power plants (Surabhi 2017).

Till recent past, fly ash is a resource material in various sectors, thus proving its worthiness over a period of time. Current efforts of ash utilization have resulted in achieving just 63.25% in 2016–17 as per Central Electricity Authority, New Delhi. It is interesting to note that the Prime Minister's office has asked for multiplying the fly ash usage by 10 times in a time-bound manner in the country to ensure clean air and mitigating adverse impact on environment. Following this directive, the Building Materials and Technology Promotion Council has proposed to make use of fly ash bricks mandatory for all central construction agencies and in the road making with fly ash in government schemes such as the Prime Minister's Awas Yojana, AMRVT and Smart Cities Mission. There is also a plan to ban on the use of clay bricks within 300 km radius of thermal power plants. The Government of India has planned to expand its scope to include bio-based ash resources. As a follow-up, National Green Tribunal has recently called for nation-wide action plan on the usage of fly ash from various states of India for promoting fly ash utilization.

It is in this context in an attempt to strengthen the mood of the Government and other agencies planning to arrive at modes of various applications of fly ash, our discovery pertaining to use of the fly ash as a pesticide and a carrier in pesticide formulation will be a valuable component in plant protection. This paper will therefore throw light on potentials of the fly ash in crop protection so as to alleviate pollution due to the chemical pesticides whose ugly eye is seen in the agricultural produces like seeds, fruits, vegetables, grains, fodder besides bringing down cost of plant protection to a considerable low level, which everyone is wanting for.

3 Potentials of Fly ash as a Pesticide

Fly ash used in our probe is mainly that of ash from lignite-based power plant like Neyveli Lignite Corporation India Limited, Neyveli, Tamil Nadu. Such lignite fly ash was subjected to grinding in ball mill to obtain fine particle size of less than 50 μm, and one fraction which is nearer to nanoparticle size was utilized in all our studies made thereafter. Findings obtained revealed an interesting picture so far as the pest control is concerned.

Key findings are illustrated hereunder.

An array of more than 100 species of insect pests damaging crops like rice, sugarcane, cotton, vegetables (brinjal, bhendi, tomato, cauliflower), fruit trees, pulses, oilseeds (Narayanasamy and Gnanakumar Daniel 1989; Nambirajan 1999). This was followed by an intensive series of research by Narayanasamy (1994) which culminated in standardizing dosages and timing of applications in rice and other crops. Further insecticidal property of the fly ash deciphered important pests like brown planthopper (BPH), green leafhopper (GLH), rice stem borer, leaf folder, grasshopper, spiny beetle, earhead bug in rice (Vijayakumar and Narayanasamy 1997), fruit borer, *Helicoverpa armigera* Hub., *Spodoptera litura* Fab. of tomato (Selvanarayanan et al. 1997), fruit borer, *Leucinodes orbonalis* Guen., *Earias vitella* Fab. (Nambirajan 1999), *Crocidolomia binotalis* Z. of cauliflower (Selvanarayanan et al. 1997) (Plate 1).

Caterpillar
Rice Stem Borer

Grub
Brinjal Leaf Beetle

Plate 1 Mortality of insects due to fly ash treatment (Narayanasamy 1994)

Table 1 Schedules of recommendation of fly ash for various crops in agriculture (Narayanasamy 2001)

Crops	Target pests	Dosages (kg/ha)	Time of application
Rice	BPH, GLH, leaf folder, caterpillars, beetles, bugs	40	30, 40, 50 and 60 DAT
Bhendi	Fruit borer, leaf eaters	40	20, 35, 50 and 65 DAS
Brinjal	Fruit borer, leaf beetle, leaf webber, grasshopper	40	35, 50, 65, 80 and 95 DAS
Tomato	Fruit borer, mealy bugs, leafhopper	40	30, 45, 50, 65 and 80 DAT

DAT days after transplanting; *DAS* days after sowing

Similar results were obtained when the fly ash was tested under field condition in various crops like rice, pulses, oilseeds, vegetables and fruit trees in a four-year research program funded by the Ministry of Coal, Government of India during 2007–2011 (Coal India Project Report 2011).

Efficacy of the fly ash against pests of rice was demonstrated in two farmer's fields in Kattumannarkoil Taluk in Cuddalore District, Tamil Nadu.

Based on the findings obtained, schedule of fly ash dust application has been formulated for adoption (Table 1).

3.1 Mode of Action of Fly ash in Insects (Narayanasamy 2001)

3.1.1 Fly ash Acts as an Insecticide

Dusting of fly ash had denuded mandibles (feeding organs) of insects like caterpillars, beetles, grasshoppers, weevils in rice, vegetables and other crops (Plate 2).

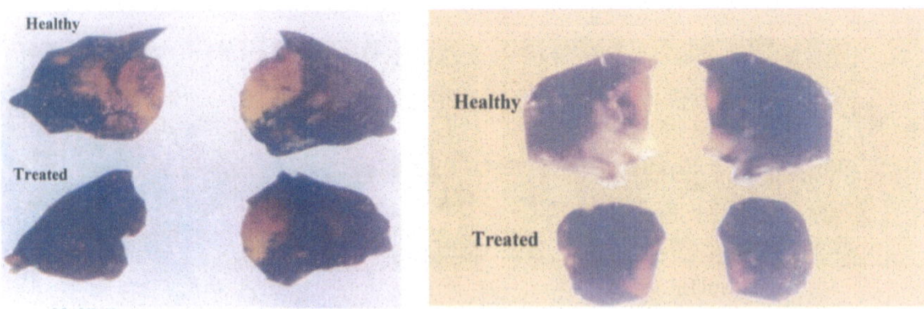

Rice Leaf folder   Rice yellow hairy caterpillar

Plate 2 Impact of fly ash on the mandibles of insects (Narayanasamy 1994)

Plate 3 Induced malformations in various insect species due to fly ash treatment (Narayanasamy 2001)

On examination of the insect cadavers dead due to fly ash, fly ash particles were accumulated in the foregut and mid gut of alimentary canal and subsequently digestion of foods impaired resulting in no feeding by the insect before succumbing to death. In cases of early consumption of fly ash treated plants by the insects in their larval stage, succeeding instars of development like pupa and adult got malformed which meant no further generation in their life (Plate 3). Fly ash particles had inhibited vital organs, enzymic secretion and growth hormones resulting in deranged growth significantly.

There were extensive damages inflicted in the tissues of the guts of *H. armigera*.

3.1.2 Fly ash Causes Loss of Body Moisture in Insects

Biochemical profiles of proteins, lipids, chitin (a major constituent of insect body wall) and body water got declined in *Catopsilia pyranthe* F. and *H. armigera* mainly due to the adsorptive and abrasive nature of the fly ash particles (Plate 4).

Secretion of enzymes like protease, amylase pectinase, chitinase and lipase were interfered with coating of fly ash on insect bodies affected loss of body moisture

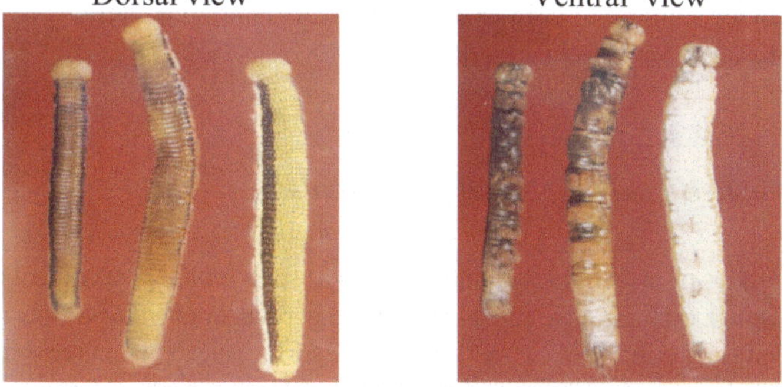

Plate 4 Adsorptive and abrasive nature of lignite fly ash on insects (Narayanasamy 1994)

through removal of lipid layer of insect body wall, before death. Fly ash nanoparticles adhered to the body surface firmly.

Hence, it is inferred that the fly ash acts not only a physical poison but also stomach poison.

3.2 Inducement of Plant Resistance Against Insect Pests

Root-zone application of fly ash in rice plants enhanced resistance factor against feeding of sucking pests, brown planthopper and green leafhopper through accumulation of silicified cells along culm and leaves of the plant, thus making the culms thick so as to act as barrier to the feeding of insects.

3.3 Safety of Fly ash to Non-target Animals

Fly ash revealed itself to be safe to albino rats, fishes and earthworms and beneficial insects like mulberry silkworms and honeybees under greenhouse conditions.

Fly ash was a good protector of beneficial fauna of rice ecosystem like spiders, predatory crickets, ladybird beetles, dragon flies and damsel flies. Besides natural incidence of entomofungal pathogens infecting various key pests like brown planthopper (BPH), leaf folder of rice was rampant in field situation.

4 Value of Fly ash as a Carrier in the Formulations of Insecticides

Agrochemical industry gives more thrust for synthesis and application of eco-friendly pesticides in agriculture so as to enable a slow and continued delivery of the agrochemicals towards the target pests.

4.1 Fly ash-Based Chemical Insecticides

Technical grades of insecticides cannot be used as such for the pest control programs as they are pure compounds, hence highly dangerous to all lives and cost higher. The formulation auxiliaries like inert materials, surfactants, solvents and few others are needed to be mixed with the technical grades. Hence, such technical grades are normally diluted into a definite proportion with suitable diluents suiting external conditions such as temperature and humidity. In this scenario, nanoparticles appear to be potentially useful as inert carrier in the present chaotic agricultural sector due to their bio-compatibility, high surface areas and other unique properties. In this context, fly ash appears promising in this nanoparticles technology now in the offing.

Certain minerals that are most frequently involved in the formulation include calcite, montmorillonite, attapulgite, magnesite and diatomite.

Suitability of flash as an inert dust was probed. Initially, insecticides like BHC 10% Dust, Malathion 25% WDP, BHC 50% WP and Carbofuran 3% Water Dispersible Granule were synthesized involving lignite fly ash as a carrier replacing the conventional carriers as per IBI standards. All these fly ash-based insecticides excelled in killing important pests of rice both under laboratory and field conditions (Narayanasamy 2001).

Recently through a project sponsored by the fly ash unit, DST, Government of India implemented during 2008–2011, four insecticides, viz. Chlorpyriphos 1.5% WDP, Endosulphan 4% Dust, Imidacloprid 2.15% Tablet and Fipronil 0.05% Tablet, were developed involving fly ash as a carrier. Of these, Chlorpyriphos and Endosulphan were found effective in checking pests of rice, brinjal and cotton, while being superior to the commercial insecticides both in the laboratory and field conditions (DST Project Report 2012).

The other products, Imidacloprid and Fipronil were tested for the domestic pest; cockroach and fipronil were found promising to kill the pest. In view of certain problems encountered, there arose a need to improve the formulation protocol for these candidate products to include more number of feeding attractants in the recipe.

Importantly, the use of fly ash as a carrier could reduce the cost of the pesticides synthesized when compared to the insecticides containing commercial carriers owing to their high cost.

4.2 Fly ash-Based Herbal Insecticides

Unlike the chemical insecticides, the plant originated compounds are comparatively safe to the mammals and higher animals besides containing the target pests in agricultural crops.

A maiden attempt was made to synthesize eight herbal insecticides involving fly ash as a carrier, viz. Fly ash + Neem seed kernel 10% D, Fly ash + Eucalyptus 10%, Fly ash + Vitex 10% D, Fly ash + Ocimum 10% D, Fly ash + Turmeric 10% D (Arputhasankari and Narayanasamy 2007).

Among the herbals synthesized, Fly ash + Turmeric 10% Dust and Fly ash + Neem seed kernel 10% Dust were found to be most effective against all the test insects including Epilachna on brinjal (Plate 1) and *S. litura* F. on bhendi followed by Fly ash + Vitex 10% D, Fly ash + Eucalyptus leaf 10% D and Fly ash + Ocimum leaf 10% Dust.

Examination of feeding organs, namely mandibles of insects treated with these products revealed that they became unfit to chew the food particles. This is due to the content of silica (as SiO_2) in the fly ash which had blunted them, with the ultimate cessation of feeding and death.

Topical application of Fly ash + Turmeric rhizome 10% D showed it to be effective against sucking pests of brinjal like lacewing bug, aphids and mealy bugs.

Besides motility of the pests tested, other effects like malformations in the emerging adults were encountered in the rice leaf folders, bhendi semilooper and bhendi fruit borer; the emerging adults, however, were short-lived.

Recently, nine fly ash-based herbal insecticides were synthesized through a four-year national project funded by the Central Mine Planning and Design Institute Ltd., Ministry of Coal, Government of India during 2007–2011 at Annamalai University, India.

They were FA + Neem 30% dust, FA + Indian Privet leaf 30% Dust, FA + Eucalyptus leaf 30% Dust FA + Pongam leaf 30% Dust, FA + Turmeric 30% Dust, FA + Chilly 30% Dust, FA + Sweet flag rhizome 30% Dust, FA + Adathoda leaf 30% Dust and FA + Dried Ginger 30% Dust. These newer pesticides were tested for pest control in rice, brinjal, bhendi, etc. Of them, FA + Dried Ginger, FA + Sweet flag, FA + Turmeric, FA + Neem and FA + Chilly revealed their potency against pests like fruit borer, leaf beetle, caterpillars and sucking pests compared to the control neemgold.

It is therefore concluded that the fly ash-based herbal pesticides are potent biopesticides and that the fly ash could be less expensive than the conventional products. The author of this paper through his explanations and ramifications on the use of the fly ash in agriculture especially in pest control given in an Interview Text Lecture programme with National Institute of Science Communication and Information Resources, New Delhi, has shown potentials of fly ash in crop protection (NISCAIR 2008).

5 Fly ash Nanoparticle Technology

A nanoparticle (or nano-powder or nano-crystalline) is a microscopic particle with at least less than 100 nm diameter. Nanoparticle research is a field of upcoming probe for application in biomedical, optical and electronic fields. Nanoparticles have more surface area to volume than the large particles. The nanoparticles have a limitation as they are tiny ones, likely to be inhaled or ingested while in use and may pose a health hazard and environmental risk. However in this fast moving global science world, this technology shows promise in purification of water in shortest possible time. In this scenario, interest arises in the generation of fly ash nanoparticles for application in agriculture.

Nano-sized particles present in coal fly ash appear to show promise in pest control strategies. TEM and SEM image revealed fly ash particle to be predominantly spherical and some polymorphic structured in the size range of 11–25 nm, while average crystalline size of the coal fly ash nanoparticles at 14 nm, lignite fly ash particles were mostly irregular and form spherical particles that were also observed (Plate 5).

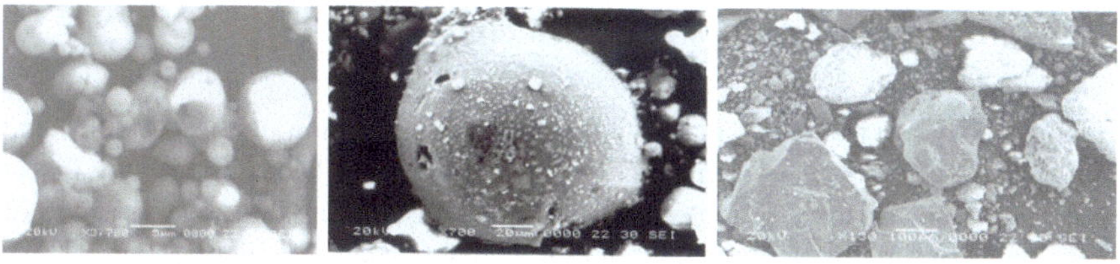

Coal fly ash — Coal fly ash Magnified — Lignite fly ash (Mag: x 130)

Plate 5 Ultra-structure of nano-fly ash particles (Narayanasamy 1994)

5.1 *Fly ash Characterization*

Fly ash particle size and morphology are of vital value in the context of invoking nanoparticle technology for possible enhancement of efficacy of fly ash in the plant protection programmes in crops.

The fly ash is characterized by its physical (lightweight, small spherical particles, hardness) and chemical properties that give it with an economic value as raw materials in many applications. Fly ash particles are classified into glassy, aluminosilicates, spongy carbonaceous, spherical metallic particles, spherical rutile particles, spherical lime particle and mineral formless.

Fly ash particles are very fine, lightweight and spherical with diameter, 1–150 μm. Fly ash is grey to blackish grey.

From elemental analysis, the fly ash was observed to be enriched in silica, alumina, calcium oxides and ferric oxides besides containing traces of Zn, Pb, Rb, Zr, Sr, S, Th, Cu, Ni, Mn, Cr, V, U, Y and Ba. Of these, silica presented as SiO_2 accounting for 80–90% in the fly ash which is the vital 'insecticidal factor' (Plate 6). Silica a well-known element followed by Al, Ca, Fe and sulphur accounts for the pesticidal property as reported by many workers. Emergence of silica-based biopesticides has been witnessed in recent times.

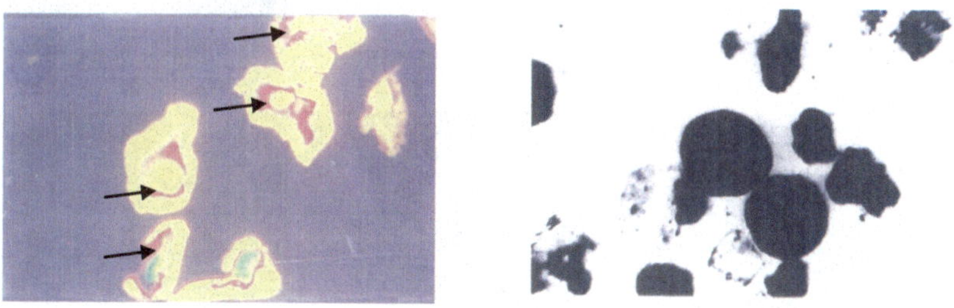

Angular shaped silica (as SiO_2) particles (100x) — Fromboidal shaped pyrite (as FeS_2) particles (100x)

Plate 6 Images of major chemical constituents of fly ash (by staining) (Narayanasamy 1994)

5.2 Nanostructures of Insect Exoskeleton

Structure of fly ash pesticides plays vital role in the deposition on insect bodies and plant surfaces considering various kinds of processes distributed in both the insect bodies and plant surface. In fact insects exhibit a fascinating and diverse range of micro- and nano-archistructures in their cuticle. Numerous cuticular structures have been discovered which imbue the cuticle with anti-wetting properties self-cleaning abilities, anti-reflection, enhanced colour, adhesion and attachment properties (Watson et al. 2017). The differently shaped particles of fly ash could adhere to the exoskeleton of the insects firmly which contain variety of structures such as hairs, scales, spine-like processes, nodules, pustules, vesicles which aid in the attachment (Hu and Watson 2011) (Plate 7). When the fly ashes were delivered in the field as dusts through mechanical dusters, they got deposited not only on the insects but also the stems and leaf culm of the crop plant, cling to them and remained for considerably long time to check pest damage.

These morpho-features of the fly ash revealed further that there was better dispersion of the filler in the synthesis of chemical and herbal insecticides while imparting carrier value to the fly ash.

These features witnessed in our series of investigations that adherence and persistence of the fly ash particles on rice leaf surface lasted for 6 days. Fineness of the lignite fly ash enhanced adherence of the dust to the body wall of insects like caterpillars, beetles with highest quantity of the fly ash adhering to the rice leaf folder larvae. Fly ash deposition was maximum at a high of 0.5 m followed by 1.0 and 1.5 m from the ground level (Narayanasamy 1994).

Scales in Butterflies | Trichome hairs in moths | Pores in Beetles

Silicified raised spots in Rice leaves | Trichome hairs in Brinjal leaves | Trichome hairs in Tomato leaves

Plate 7 Semi-ultrastructures in insect body cuticle and crop plants surface (Narayanasamy 2012)

5.3 Crop Plant Surface Morphology

There is diversity of plant surface structures ranging from single cell to multicellular surface sculptures. An array of structures detected includes hairs, wax crystals and surface folding play a special role. It is hoped that this new approach might stimulate those concerned to initiate or intensify the study of biological surface in respect of fly ash dusting. Morphological traits of rice plant such as, plant culm height, closer-spacing, tiller number per hill, tiller perimeter, leaf size, and number, leaf area index and open canopy structure are of value both in the deposition of fly ash particles and also distribution of the insects which determine length of insect feeding tenure and quantum of food intake.

Rice plant possess a pair of ear-like appendages called ‘auricles’ found on either side of the base of the blade and another appendage, ‘ligule’, a papery membrane at the inside juncture between the leaf sheath and leaf lamina. It is felt that these structures might aid in accumulation of fly ash particles at these spots when dusted to tackle crawling insects like caterpillars, maggots.

Further, rice plants have rows of sclerenchymatous cells in the culm, leaf sheath, leaf lamina containing lignin, impregnated with silica which normally injures the feeding organs of sucking and chewing pests. Similarly, brinjal plant was borne with abundant multi branched trichome hairs and tomato plants with four types of trichome hairs which not only hold the fly ash particles and also inhibit locomotion of neonate larvae warranting their early mortality (Plate 7).

So far as *H. armigera* is concerned, the adult moths preferred broader leaves of crops like bhendi, tomato, maize for ovipositor their eggs and on brinjal flowers and fruits were preferred, leaves of bhendi were the preferred feeding sites for neonate larvae.

From the above, it is evident that micro- and nanostructures present in the plant’s anatomy and insect exoskeleton indicates their suitability to application of fly ash nanoparticles technology for highly selective type of pest control programmes, through detailed probe in this regard is necessary.

6 Conclusions

Findings pertaining to pesticidal and carrier values of the fly ash made so far are exhaustive and conclusive, thus paving way for commercial exploitation.

Standardization technique to generate cost-effective nanoparticles of fly ash need attention for advanced utilization of the ash in agriculture.

Use of fly ash nanoparticles in the synthesis of pesticide formulations appear interesting need further exploration.

Let me quote the national address of our former President of India, Dr. A. P. J. Abdul Kalam promoting the use of the fly ash on the eve of the country’s 56th Republic Day.

Use of coal for power generation has resulted in large quantity of fly ash touching 100 mill. tonnes/year. We need to find ways to utilize this ash in various ways. Agricultural production of grains is around 15% greens, 35% vegetables and root tubers 50% when the fly ash is mixed in soils. Toxicity studies have proved that there is no toxic element due to fly ash. Nutritive values were increased due to availability of iron and calcium. The fly ash thus can become a wealth generator by making use of it for producing 'green' building materials, roads, agriculture, etc.

We feel proud herein to be a front runner to enlarge the applications of the fly ash in the form of various newer fly ash-based chemical and herbal pesticides towards promoting green agriculture while enabling the bulk disposal of the fly ash at the premises of thermal power plants.

Acknowledgements The author is grateful for the financial supports provided by Tamil Nadu State Council for Science and Technology, Government of Tamil Nadu, Chennai, Tamil Nadu, Fly ash Unit, Department of Science and Technology, Government of India, New Delhi, and Central Mine Planning Design Institute Ltd., Dhanbad of Ministry of Coal, Government of India, New Delhi, India, for undertaking these maiden research projects. Further, my grateful regards are due to authorities of Annamalai University, Tamil Nadu, India, for permission and provision of infra-facilities and manpower to organize and complete all the three projects in time. Personally, I am much grateful to Dr. Vimal Kumar, Former Head, Fly ash Unit, Department of Science and Technology, Government of India, New Delhi, for his sustained support and co-operation in the grant of the project.

References

Arputhasankari, S., & Narayanasamy, P. (2007). Bio-efficacy of flyash based herbal pesticides against pests of rice and vegetables. *Current Science, 92*(6), 811–816.

Coal India Project Report. (2011). Project completion report on development and use of flyash-based pesticides funded by Ministry of Coal, Government of India, New Delhi.

DST Project Report. (2012). Final project report on use of flyash as a carrier in pesticides formulations funded by Flyash Unit, DST, Government of India.

Hu, H.-M., & Watson, J. A. (2011). Fouling of nano-structured insect cuticle: Adhesion of natural and artificial contaminants. *The Journal of Bioadhesion and Bioflim Research, 27,* 1125–1137.

Nambirajan, S. G. (1999). *Further studies on the use of flyash 100% dust against key pests of certain vegetable crops*. M.Sc. (Ag.) Thesis. Annamalai University, Annamalai Nagar, Tamil Nadu, India.

Narayanasamy, P., & Gnanakumar Daniel, A. (1989). Lignite flyash: A non-polluting substance for tackling pest problems. In K. V. Devaraj (Ed.), *Progress in pollution research* (pp. 201–206). Hebbal, Bengaluru, India: University of Agricultural Sciences.

Narayanasamy, P. (1994). Final report of Scheme titled studies on use of lignite flyash as an insecticide and an adjuvant in insecticide formulations supported by the Tamil Nadu State Council for Science and Technology, Government of Tamil Nadu, Adayar Chennai, India, 99p.

Narayanasamy, P. (2001). Flyash pesticides in the millennium. In P. Narayanasamy (Ed.), *Agricultural Application of Flyash, Proceedings of the National Seminar on Flyash* (pp. 1–6). Annamalai University, Annamalai Nagar, Tamil Nadu, India.

Narayanasamy, P. (2012). *Exploiting particle size, morphology and mineralogy of flyash for enhanced pest control in agriculture*. Paper presented at National Seminar on Prospects of Flyash as a Mineral Resource, Organized by CSIR-Institute Minerals and Materials Technology, Government of India, Bhubaneswar, held in July 17–18, 2012, pp. 24–25.

NISCAIR. (2008). Transcript of interview with Dr. P. Narayanasamy for use of fly in pest control. In Planet earth: The road ahead (pp. 177–181). National Institute of Science Communication and Information Resources (NISCAIR), Pusa, New Delhi, India.

Selvanarayanan, V., Ganesean, K., & Narayanasamy, P. (1997). Insecticidal activity of lignite flyash against the larvae of *Spodoptera litura* Fab and *Crocidolomia binotalis* Zeller. In P. Narayanasamy (Ed.), *Flyash in agriculture* (pp. 8–11). Annamalai Nagar, Tamil Nadu, India: Annamalai University.

Surabhi. (2017). Flyash in India: Generation vis-à-vis utilization and global perspective. *International Journal of Applied Chemistry,* 13(1): 29–52.

Vijayakumar, N., & Narayanasamy, P. (1997). Use of flyash as an inert carrier in insecticide formulations. In P. Narayanasamy (Ed.), *Flyash in agriculture* (pp. 31–33). India: Annamalai University, Annamalainagar, Tamil Nadu.

Watson, G. S., Watson, J. A., & Cribb, B. W. (2017). Diversity of cuticular micro-and nanostructures on insects: Properties, functions and potential applications. *Annual Review of Entomology, 62,* 185–205.

Behaviour of Fly Ash Concrete at High Temperatures

A. Venkateswara Rao and K. Srinivasa Rao

Abstract The present study deals with the behaviour of fly ash concrete at elevated temperatures. Concrete specimens of cubes of 100 mm, cylinders of 100 mm diameter and 300 mm height and prisms of 100 × 100 × 500 mm are cast by replacing the cement with fly ash in a range of 30–50% by mass of cement. The specimens are cured for 7 and 28 days, then dried and are exposed to elevated temperatures ranging from 100 to 500 °C for a period of 1 and 3 h, to investigate the effect of temperature on compressive strength, split tensile strength and flexural strength. It is observed that the residual compressive strength, split tensile strength and flexural strength of fly ash concrete initially increased with increase in temperature, later decreased with further increase in temperature.

Keywords Fly ash · Compressive strength · Split tensile strength · Flexural strength elevated temperatures

1 Introduction

Concrete is the most extensively used and preferred construction material in development of infrastructure due to its added advantages such as easily mouldable to any shape and sustainability in aggressive environment. In concrete, cement is the binder and also responsible for strength. Manufacturing of cement involves in consuming lot of energy and emitting a lot of CO_2 causing the environmental problem and also the cost of the cement is high. Hence, an alternative material to cement is required to save energy and environment. Previous investigations reveal that concrete has been prepared by introducing some additives called mineral admixtures,

A. Venkateswara Rao (✉)
Department of Civil Engineering, KL Deemed to be University, Vijayawada, Andhra Pradesh 522 502, India
e-mail: avenkat424@gmail.com

K. Srinivasa Rao
Department of Civil Engineering, College of Engineering, Andhra University, Visakhapatnam, Andhra Pradesh 530 003, India
e-mail: killamsetti@yahoo.com

S. K. Ghosh and V. Kumar (eds.), *Circular Economy and Fly Ash Management*,
https://doi.org/10.1007/978-981-15-0014-5_8

beside its basic ingredients to investigate their effect on strength, durability and other properties of concrete. Fly ash is one such material which is the waste product in thermal plants and easily available at low cost, if not suitably disposed off creates environmental problems. Past investigations have revealed that fly ash can be used as an important constituent of concrete (Malhotra and Mehta 2005; Hand Book on High volume fly ash concrete technology 2005). It can be used as either an admixture or as partial replacement of cement or replacement of fine aggregate. Concrete may be exposed to high temperatures in nuclear reactors, oil wells, pavements of air craft, fire accidents in buildings, etc. In such situations, concrete should give better performance. In general, when concrete exposed to elevated temperatures, it undergoes many microstructural changes and affects performance of structure. So, the present study involves in replacement of cement with different percentages of fly ash and its behaviour at different elevated temperatures.

2 Previous Studies

Raju and Rao (2001) studied the effect of high temperature on compressive strength of concrete by replacing cement partially with fly ash in range 10–30% at temperatures, of 200–250 °C for 1, 2 and 3 h duration of exposure, and an increase in compressive strength of concrete was reported.

Sarshar and Khoury (1993) tested ordinary Portland cement (OPC)—pulverized fly ash (PFA) paste containing 30% fly ash by weight of cement at various temperatures up to 650 °C. It was reported that the residual compressive strength of fly ash concrete was found to be 88% at 450 °C and 73% at 600 °C.

Khan et al. (2013) prepared specimens of three mixes replacing cement of 43 grade with 40, 50 and 60% fly ash by weight. The specimens after curing for 28 days are exposed to temperature ranging from 100 to 900 °C. From the study, it was reported that, in all mixes, there is an increase in compressive strength up to 300 °C with increase in temperature and with further increase in temperature, the compressive strength is decreased. Also, observed that for same fly ash content (50%), the drop in compressive strength is more in case of specimens with lower water to binder ratio than that with more water to binder ratio.

Srinivasa Rao et al. (2006) have studied M20, M30 grade concrete using OPC of 53 grade, fly ash from Vijayawada Thermal Power Station, Vijayawada and exposing specimens to thermo-cycles at different temperatures and evaluated the compressive strength, split tensile strength and dynamic modulus of elasticity. It was reported that the compressive strength, split tensile strength and dynamic modulus of elasticity were decreased for specimens of OPC, while there is increase in them for specimens with fly ash. It was reported from the results that the concrete with fly ash is more effective in resisting thermo-cycles compared to concrete with OPC.

Diederichs et al. (1989) tested ASTM class F fly ash concrete specimens of 90 MPa exposing them to temperature range of 200–250 °C and observed that there is an increase in strength as the temperature increases.

Potha Raju et al. (2004) carried out experiments on M28, M33 and M35 grades of concrete, investigated the effect of temperature on the flexural strength of fly ash concrete by partial replacing cement with fly ash (10, 20 and 30%) and heated to 100, 200 and 250 °C for 1, 2 and 3 h duration It was reported that fly ash concrete showed consistently the same pattern of behaviour as that of concrete without fly ash under elevated temperatures during flexure. The fly ash concrete with fly ash content up to 20% showed better performance compared with the specimens without fly ash by retaining a greater amount of its strength.

3 Experimental Program

3.1 Mix Proportions

In the present experimental investigation on concrete, cement is replaced with varying fly ash content of 30, 40 and 50% by weight of cement besides one mix without fly ash, total four mixes are selected. The mix proportion for the present study is 1:1.39:3.43 (cement:fly ash:fine aggregate:coarse aggregate) arrived after several trial mixes from guide lines of IS:10262-2009 and IS:456-2000. The ordinary Portland cement of 53 grade (IS:12269), fine aggregates conforming to zone-II (IS:383-1970), coarse aggregate as per IS 383 code provisions and class F fly ash are used for present study.

3.2 Method of Exposure and Testing

The cubes of 100 mm size are cast for studying the response of different concrete mixes at elevated temperatures. The specimens are demoulded after 24 h from time of casting and placed in potable water for curing. After specified period (7 and 28 days) of curing, the cubes are removed from water and allowed to dry in air before exposing to elevated temperatures. Three control specimen cubes of 100 mm size are tested at ages of 7 and 28 days for each percentage and age, and the average value of compressive strength is considered. Weights of specimens are measured with digital balance before and after heating are recorded.

An electric furnace with a maximum operating temperature of 1050 °C is used to heat the specimens. Inside temperature of the furnace is at room temperature at the time of placing the specimens in furnace. The temperature of furnace as set with help of digital temperature control panel board (PID controller) of the furnace. After reaching the desired temperature, it is kept constant throughout the duration of exposure (1 h or 3 h).

As the specified duration of exposure is completed, the furnace is switched off and the specimens are allowed to cool to room temperature in air (air cooling) and tested for compression (IS:516-1959), in 3000 kN compression testing machine and

Fig. 1 Cast of prisms on table vibrator

Fig. 2 Specimens arranged for heating in furnace

for split tension (Short et al. 2001) and flexure in 10 MT universal testing machine. Photographs of the cast specimens are shown in Fig. 1 and arrangement of specimens for heating in furnace is shown in Fig. 2.

4 Results and Discussions

4.1 Residual Compressive Strength

Residual compressive strength is the ratio of compressive strength of heated specimen of any percentage of fly ash or age to that of 28 day strength of controlled concrete (reference concrete) at room temperature multiplied by 100. The percentage residual compressive strength at age of 7 days and 28 days are shown in Table 1. Figures 3, 4, 5, 6, 7, 8, 9 and 10 show the graphical representation.

At the age of 7 days, the maximum residual compressive strength of 94.5% is exhibited by SC without fly ash at 200 °C and 1 h exposure duration while minimum 54.6% is exhibited by SC with 50% fly ash at 500 °C and 3 h duration of exposure. At age of 28 days, the maximum residual compressive strength of 105.3% is exhibited by SC with 30% fly ash at 300 °C and 1 h exposure duration while minimum 66.5% is exhibited by SC with 50% fly ash at 500 °C and 3 h duration of exposure.

Khan et al. (2013) have also reported similar trend. Diedrichs et al. (1989) tested fly ash concrete in a temperature range of 200–250 °C and reported that there was an increase in strength of concrete with an increase in temperature as observed in this study.

4.2 Residual Split Tensile Strength

Residual split tensile strength is the ratio of split tensile strength of heated specimen of any percentage of fly ash or age to that of 28 day strength of controlled concrete (reference concrete) at room temperature multiplied by 100 (Table 2).

4.3 Residual Flexural Strength

Residual flexural strength is the ratio of flexural strength of heated specimen of any percentage of fly ash or age to that of 28 day strength of controlled concrete (reference concrete) at room temperature multiplied by 100. It is observed that at age of 7 days, maximum residual flexural strength of 88.9% is exhibited by SC without fly ash at 200 °C and 1 h exposure duration while minimum 44.4% is exhibited by SC with 50% fly ash at 500 °C and 3 h duration of exposure. While at age of 28 days, the maximum residual flexural strength of 104% is exhibited by SC with 30% fly ash at 300 °C and 1 h exposure duration while minimum 57.6% is exhibited by SC with 50% fly ash at 500 °C and 3 h duration of exposure (Table 3; Figs. 11, 12, 13 and 14).

4.4 Loss of Weight

The results of the percentage loss in weight of specimens exposed to 100, 200, 300, 400 and 500 °C at an age of 7, 28, 56 and 91 reveal that the percentage loss of weight of concrete specimens increased with increase in the temperature. For a given temperature, the percentage loss of weight increased with increase in fly ash content as well as increase in duration of exposure.

Table 1 Percentage residual compressive strength at various temperatures

S. No.	Age in days	Temperature (°C)	Percentage replacement of fly ash							
			0		30		40		50	
		Duration of exposure	1 h	3 h	1 h	3 h	1 h	3 h	1 h	3 h
1	7	27	90.2		87.9		85.4		77.2	
2	7	100	94	92.5	89.1	89	84.6	84.1	73.5	71.8
3	7	200	94.5	92.8	89.9	89.3	83.7	80.8	72.5	69.3
4	7	300	93	91.6	90.6	89.5	82.7	78.1	71.8	67.5
5	7	400	82.9	81.1	80	77.1	75.3	66.6	66.4	63.7
6	7	500	79.5	74.3	78.2	63.6	65.7	57.1	55	54.6
7	28	27	100		102.2		89.6		79.3	
8	28	100	102	101	103	102.4	87.8	86.8	75.6	75.2
9	28	200	102.6	101.6	104.8	103	85.5	84.8	74.4	73.9
10	28	300	101.2	100.3	105.3	104.1	83.1	80.3	72.6	71.7
11	28	400	101.2	93.6	102.8	98.8	80.9	78.2	69.3	68.9
12	28	500	90.5	73.7	96.9	81.2	78.4	75.2	67.2	66.5

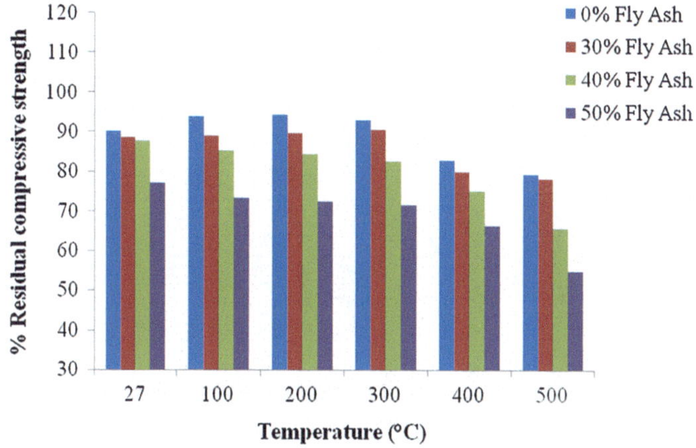

Fig. 3 Variation of percentage residual compressive strength with temperature for M30 grade concrete for 7 days at 1 h exposure duration

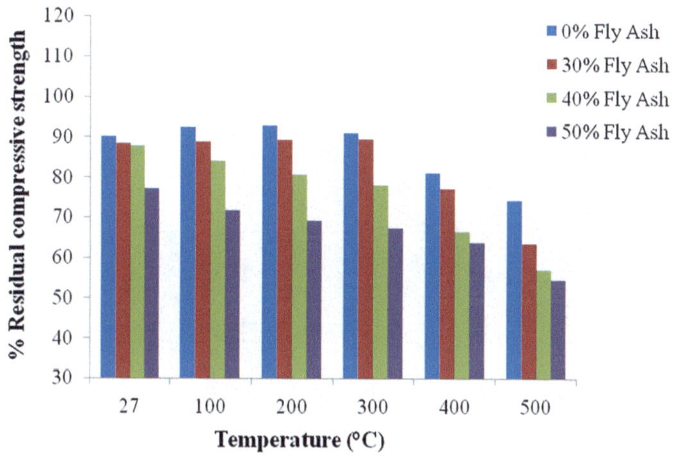

Fig. 4 Variation of percentage residual compressive strength with temperature for M30 grade concrete for 7 days at 3 h exposure duration

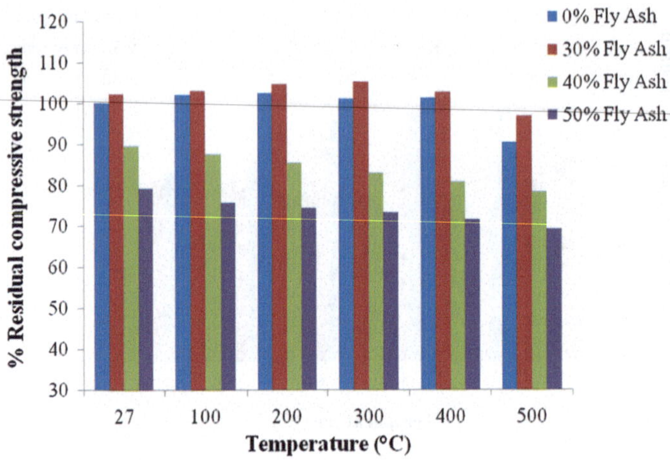

Fig. 5 Variation of percentage residual compressive strength with temperature for M30 grade concrete for 28 days at 1 h exposure duration

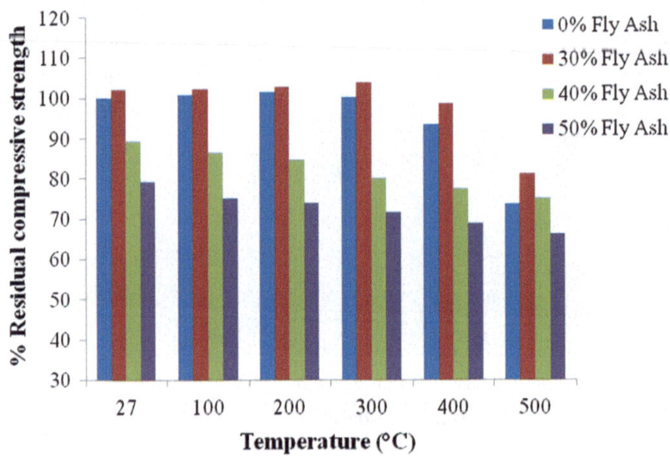

Fig. 6 Variation of percentage residual compressive strength with temperature for M30 grade concrete for 28 days at 3 h exposure duration

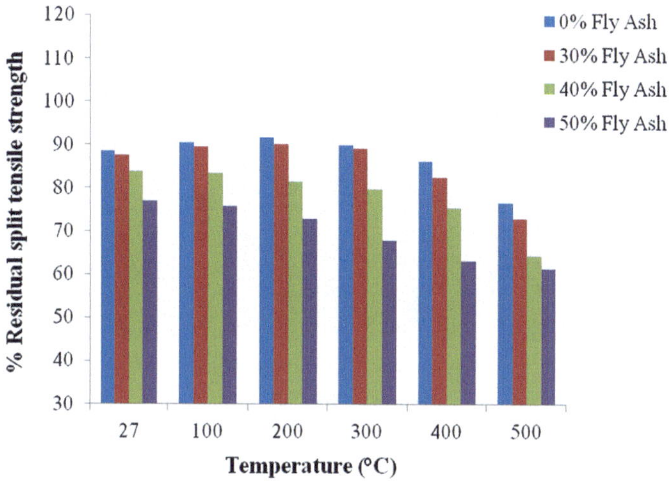

Fig. 7 Variation of percentage residual split tensile strength with temperature for M30 grade concrete for 7 days at 1 h exposure duration

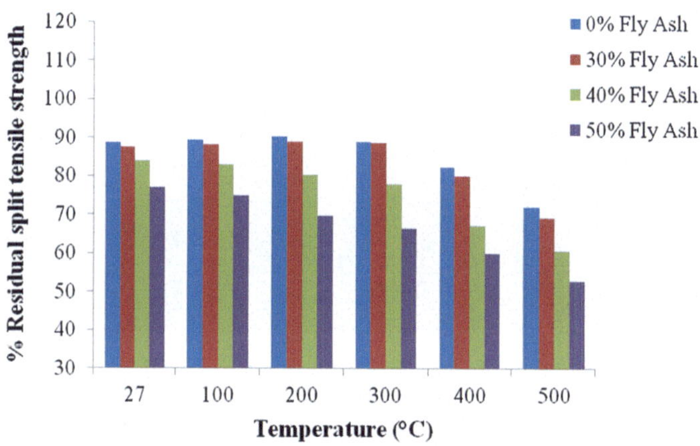

Fig. 8 Variation of percentage residual split tensile strength with temperature for M30 grade concrete for 7 days at 3 h exposure duration

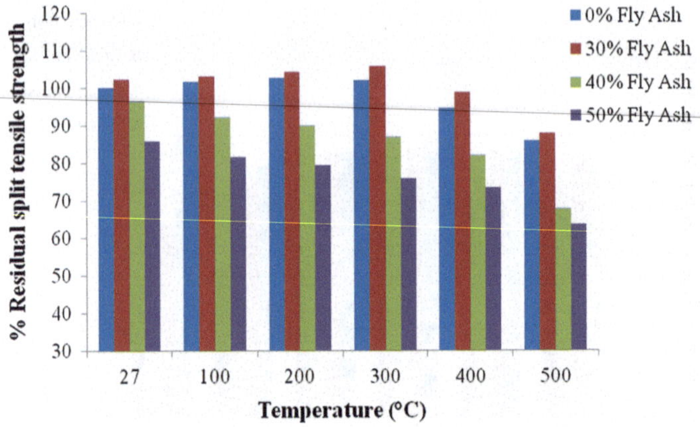

Fig. 9 Variation of percentage residual split tensile strength with temperature for M30 grade concrete for 28 days at 1 h exposure duration

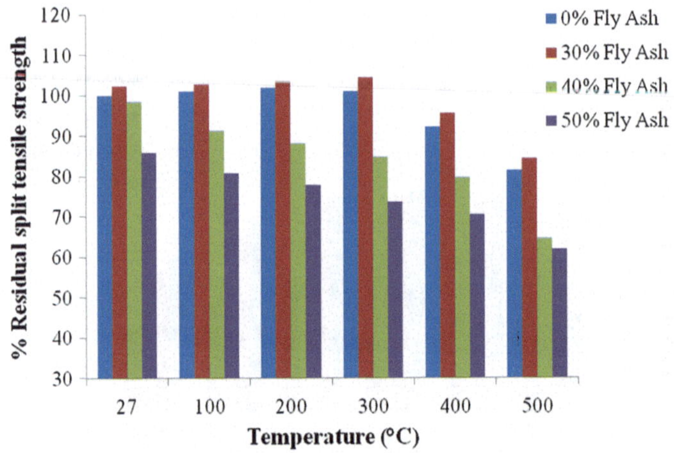

Fig. 10 Variation of percentage residual split tensile strength with temperature for M30 grade concrete for 28 days at 3 h exposure duration

Table 2 Percentage residual split tensile strength at various temperatures

S. No.	Age in days	Temperature (°C)	Percentage replacement of fly ash							
			0		30		40		50	
		Duration of exposure	1 h	3 h	1 h	3 h	1 h	3 h	1 h	3 h
1	7	27	88.7		87.7		84		77.1	
2	7	100	90.1	89.4	89.5	88.3	83.4	83	75.8	74.9
3	7	200	91.3	90.5	90.1	89	81.6	80.2	***73***	69.8
4	7	300	90	89	89.2	89.5	79.8	78	67.9	66.6
5	7	400	86.3	82.2	82.6	80	75.5	67.3	63.1	60
6	7	500	76.7	72	73	79.3	64.3	60.7	61.3	53
7	28	27	100		102.4		96.5		85.9	
8	28	100	101.5	101.5	103.2	103.7	92.4	92	81.6	81
9	28	200	102.7	102.5	104.2	104.5	90	89.2	79.5	78
10	28	300	102	101.9	105.9	105	86.9	85.9	75.9	74.6
11	28	400	94.6	93.9	98.7	96.6	82.2	80.8	73.4	70.7
12	28	500	85.7	84.9	87.9	85.6	67.9	64.7	63.7	61.9

Table 3 Percentage residual flexural strength at various temperatures

S. No.	Age in days	Temperature °C	Percentage replacement of fly ash							
			0		30		40		50	
		Duration of exposure	1 h	3 h	1 h	3 h	1 h	3 h	1 h	3 h
1	7	27	87.6		82.5		78.7		74.6	
2	7	100	88.3	87,9	83.7	83	76.5	76.2	70.7	70.1
3	7	200	88.9	88.4	84.8	84.2	75.6	74.8	68.9	68
4	7	300	87.9	87.4	85.6	85	71.6	70.1	65.8	66.9
5	7	400	75.7	72.4	72.9	70.1	69.5	68.5	60.6	58.7
6	7	500	65.6	61.6	61.7	55	55	48.2	47.9	44.4
7	28	27	100		101		94.3		92.4	
8	28	100	101.2	100.6	102	101.5	92.6	91.6	90.7	88.2
9	28	200	101.9	101	102.8	102	85.7	83.4	80.6	78.9
10	28	300	101	100.4	104	103	80.5	77.9	70,7	75.6
11	28	400	96.7	94.2	98.5	96.6	75.5	72.9	65.6	63.2
12	28	500	90.2	89	92.8	89.8	67.6	62.9	62.7	57.6

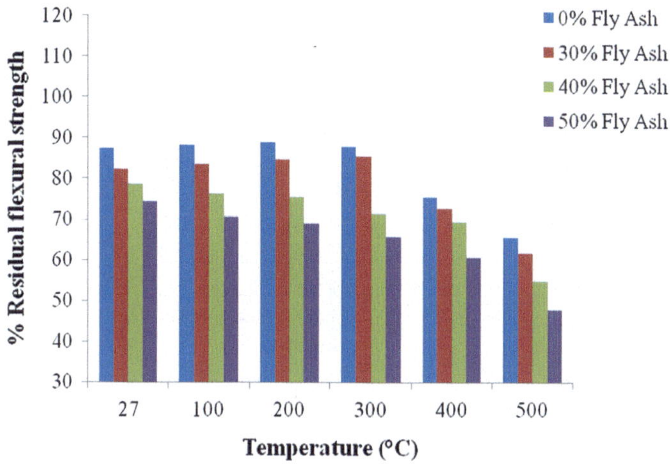

Fig. 11 Variation of percentage residual flexural strength with temperature for M30 grade concrete for 7 days at 1 h exposure duration

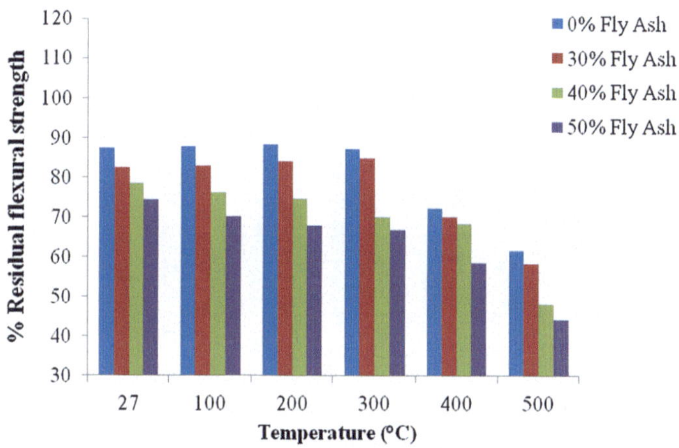

Fig. 12 Variation of percentage residual flexural strength with temperature for M30 grade concrete for 7 days at 3 h exposure duration

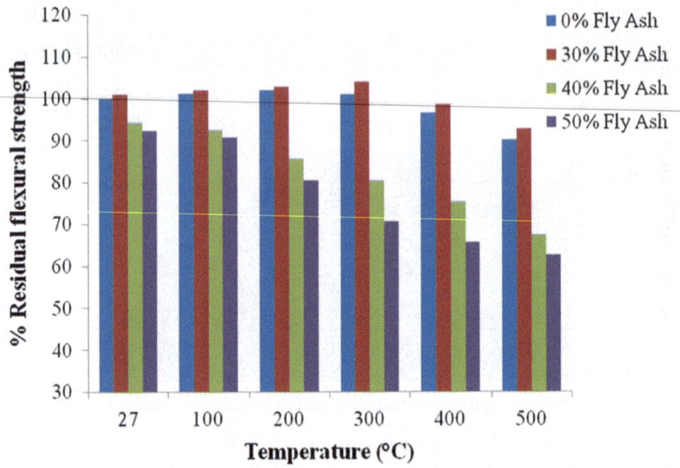

Fig. 13 Variation of percentage residual flexural strength with temperature for M30 grade concrete for 28 days at 1 h exposure duration

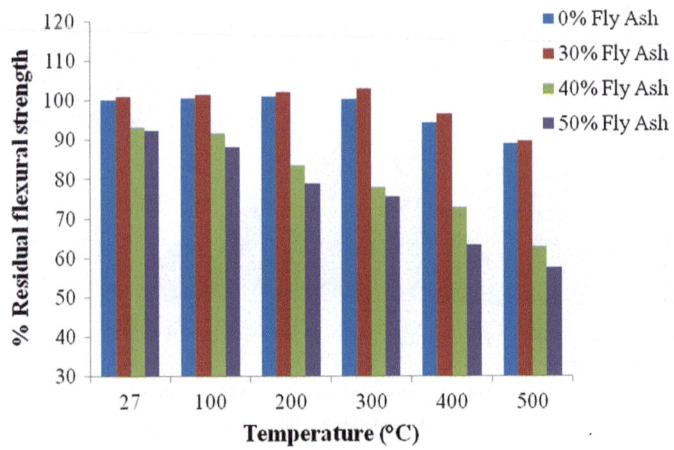

Fig. 14 Variation of percentage residual flexural strength with temperature for M30 grade concrete for 28 days at 3 h exposure duration

5 Conclusions

The following conclusions are drawn from the present experimental investigation on the behaviour of fly ash concrete at different temperatures.

1. The residual compressive strength, residual split tensile strength and flexural strength of fly ash concrete increased initially with increase in temperature up to 300 °C.
2. At a particular temperature, the percentage residual compressive strength increased for concrete with 30% fly ash, while the values decreased for 40 and 50% fly ash compared to that for concrete without fly ash.
3. At the age of 7 days, the values of residual compressive strength, split tensile strength and flexural strength for mixes containing fly ash are less than respective values for concrete without fly ash at all temperatures for both 1–3 h durations of exposure.
4. The percentage loss of weight of concrete specimens increased with increase in the temperature as well as duration of exposure.
5. For a given temperature, the percentage loss of weight increased with increase in fly ash content.
6. The deterioration of concrete increased with increase in temperature. No crack formation is observed up to the temperature of 500 °C.
7. At early ages, compressive strength of fly ash concrete is less than that of concrete without fly ash, and at later ages, compressive strength of fly ash concrete is better.
8. Finally, from the present investigation, it is concluded that replacement of cement up to 30% of fly ash in concrete is preferable for design of structures exposed to temperatures up to 300 °C.

References

Diedrichs, U., et al. (1989). *Behavior of high strength concrete at high temperatures* (Report No.92). Department of Structural Engineering, Helsiniki University of Technology.

Hand Book on *High volume fly ash concrete technology*. (2005). (3rd ed.).

IS:10262-2009, Recommend guidelines for concrete mix design. New Delhi: Bureau of Indian Standards.

IS:12269, Specifications for 53 grade portland cement. New Delhi, India: Bureau of Indian Standards.

IS:383-1970, Specifications for coarse and fine aggregates from natural sources for concrete. New Delhi, India: Bureau of Indian Standards.

IS:456-2000, Indian code of practice for plain and reinforced concrete for general building construction. New Delhi, Bureau of Indian Standards.

IS:516-1959 (Reaffirmed 1999), Indian standard methods 0f tests for strength of concrete. New Delhi, India: Bureau of Indian Standards.

Khan, M. S., et al. (2013). Effect of high temperature on high volume fly ash concrete. *Springer, Arab Journal Science Engineering, 38,* 1369–1378.

Malhotra, V. M., & Mehta, P. K. (2005). *High performance high volume fly ash concrete*. Ottawa, Canada: Supplementary Cementary Materials for Sustainable Development Inc.
Potha Raju, M., Shobha, M., & Rambabu, K. (2004). Flexural strength of fly ash concrete under elevated temperatures. *Magazine of Concrete Research, 56*(2), 83–88.
Raju, M. P., & Rao, A. J. (2001). Effect of temperature on residual compressive strength of fly ash concrete. *Indian Concrete Journal, 75*(5), 347–351.
Sarshar, R., & Khoury, G. A. (1993). Material and environmental factors influencing the compressive strength of unsealed cement past and concrete at high temperatures. *Concrete Research Magazine, 45*(162), 51–60.
Short, N. R., Purkiss, J. A., & Guise, S. E. (2001). Assessment of fire damaged concrete using colour image analysis. *Construction and Building Materials, 15*(1), 9–15.
Srinivasa Rao, P., Sravana, P., & Seshagiri Rao, M. V. (2006). Effect of thermal cycles on the strength properties of OPC and fly ash concretes. *The Indian Concrete Journal, 80*(3), 49–52.

Effect of Fly Ash on Strength of Concrete

A. Venkateswara Rao and K. Srinivasa Rao

Abstract An enormous amount of fly ash is generated in thermal power plants in India, which affects the environment and living organism. By means of fly ash in concrete in place of cement reduces the consumption of natural resources, diminishes the environmental pollution and also economical as replacing the cement which is costlier by the fly ash. The recent investigations by many researchers have revealed that the use of cementitious materials such as fly ash and GGBS in concrete is trustworthy and also economical. This present experimental research is to study the using non-conventional building material like fly ash for the development of new material and technologies. The present study involves replacing of cement with fly ash in the range of 30, 40 and 50% by mass of cement for M-30 grade concrete mix with 0.48 water cementitious material ratio. Concrete cubes, cylinders and prisms are cast, tested and compared in terms of strength at different ages of curing. The experimental results have revealed that the compressive strength has increased marginally for specimens with replacement of cement by 30% fly ash at age of 7, 28, 56 and 91 days.

Keywords Fly ash · Compressive strength · Concrete · Split tensile strength · Flexural strength

1 Introduction

Concrete is preferred and extensively used as construction material in the development of infrastructure due to its added advantages such as easily mouldable to any shape and sustainability in aggressive environment. Cement, fine and coarse aggregate and water are the main ingredients of concrete. In concrete, cement is the binder and responsible for the strength of concrete and costs more. Production of cement

A. Venkateswara Rao (✉)
Civil Engineering Department, Koneru Lakshmaiah Education Foundation, Vaddeswaram, Andhra Pradesh 522 502, India
e-mail: avenkat424@gmail.com

K. Srinivasa Rao
Civil Engineering, College of Engineering, Andhra University, Visakhapatnam, Andhra Pradesh 530 003, India

S. K. Ghosh and V. Kumar (eds.), *Circular Economy and Fly Ash Management*,
https://doi.org/10.1007/978-981-15-0014-5_9

involves in consuming more energy, emitting vast amount of CO_2 causing environmental problems and hence an alternative material that saves energy and environment is required. Previous investigations reveal that concrete can be prepared by introducing some additives, called mineral admixtures beside its fundamental ingredients to explore their effect on durability, strength and other properties of concrete. One such admixture is fly ash, a waste product in thermal power plants also easily available if not properly disposed off, creates the environmental problems and is available at low cost. Past investigations (Malhotra and Mehta 2005; Hand Book on High volume fly ash concrete technology 2005) have revealed that fly ash can be used as a chief ingredient of concrete (Malhotra and Mehta 2005; Hand Book on High volume fly ash concrete technology 2005) and can be used as either an admixture or as partial replacing cement or replacing fine aggregate partially. Therefore, the present study involves in replacing cement with different percentages of fly ash by mass and investigating its behaviour at different ages of curing.

2 Previous Studies

Khan and Prasad (2015) have studied the effect of elevated temperatures on the cube strength in and cylindrical strength in split tension of high volume fly ash concrete replacing fly ash in the range of 40–60% by the mass of cement. Specimens of 10 cm diameter and 20 cm height are cast, exposed to temperatures between 100 and 900 °C and tested for compression and split tension at an age of 28 days of curing. Results have revealed; the cube strength of concrete in compression improved initially with the increase in temperature between 200 and 300 °C; concrete strength in split tension was also improved with the increase in temperature and exhibited a maximum value at 300 °C. Further hike in temperature causes in decrease in both the cube compressive strength as well as split tensile strength.

Siddique and Khatib (2010) have studied the mechanical properties and abrasion resistance concrete containing high volume fly ash by replacing sand by class F fly ash at 35, 45 and 55% of by mass of cement. The water/binder ratio and workability of the concrete mix are maintained constant at 0.46 and 55 mm, respectively. The mechanical properties are expressed as a depth of wear. From the values at age of 28 days, it has revealed that the replacing of sand with fly ash shows the improved values, depending upon the content of fly ash and the improvement in mechanical properties continued up to the ages of 365 days.

Pandhye and Doe (2016) have explained that by reducing the fly ash mixture ratio, the cube strength of concrete in compression was increased, and replacing fly ash to 40% was better. With the increase in the content of fly ash, there is a sharp improvement in strength with change of age from 7 to 28 days signifying that at early ages concrete strength is low, also reported that for M50 concrete by 60% fly ash shows greater concrete strength at an age of 28 days. While for M30 grade fly ash concrete replacing cement by 40% of fly ash shows better strength compared to concrete without fly ash.

Basha et al. (2014) have used fly ash as a mineral admixture and assessed the cube strength in compression. Two mixes of M25 and M30 grade are designed as per stipulations of code IS-10262-82 replacing cement by fly ash in the range of 0–40% with 10% increment. Cubes of size 150 × 150 × 150 mm were cast and tested for compression at ages of 7, 14, 21 and 28 days. Results are compared to those values of conventional concrete, in these mixes increase in the percentage of fly ash to decrease the values of compressive strength.

Okoye and Singh (2016) have designed five concrete mixes to determine the effect of fly ash on workability, compressive, tensile and flexural strengths of concrete. Portland cement by weight was replaced with 20, 30, 40 and 60% fly ash, respectively, while 100% OPC was used for controlled concrete. The mechanical properties were determined at the ages of 3, 14, 28, 56 and 90 days of hydration. It was reported that by increasing fly ash, the rate of heat of hydration is decreased. X-ray diffraction and SEM studies have shown the formation of different hydration products. The concrete containing different proportions of fly ash showed better workability compared to conventional concrete. Higher the replacement value of cement by fly ash, the higher was the workability. With the increase in fly ash content, compressive strength of concrete decreased, and the compressive strengths obtained were of the order that could be used for normal construction work. The flexural strength decreased with increase in the proportion of fly ash. Finally, it was concluded as concrete with high volume fly ash can be used in normal construction work.

3 Experimental Work

3.1 Properties of Material

The properties of material used for the present work are presented below (Table 1):

Cement: OPC 53 grade conforming to IS 12269
Fly ash: Fly ash having specific gravity of 2.1, collected at N.T.P.C. Visakhapatnam, Andhra Pradesh, India.
Water: Locally available potable water with pH value 7.65 conforming to IS 3025-1986 specifications.
Fine aggregate: Locally existing river sand conforming to IS 383 with specific gravity of 2.59 and fineness modulus of 2.79.
Coarse aggregate: Blue granite stone, 2/3 of it passing through sieve of 20 mm of size and retained in sieve of 10 mm size and 1/3 of it passing through 10 mm sieve of size and retained in 4.75 mm size sieve conforming to IS 383 with specific gravity of 2.77 and fineness modulus of 6.67.

Table 1 Physical properties of cement

Particulars	Experimental values	Specifications of IS:12269-2013
Specific gravity	3.13	
Physical requirements		
1. Fineness (m^2/kg)	282	Minimum 225
2. Setting time (minutes)		
(a) Initial (b) Final	194 278	Minimum 30 Maximum 600
3. Soundness:		
(a) Le-chatelier expansion (mm) (b) Autoclave (%)	1.00 0.02	Maximum 10 Maximum 0.8
4. Compressive strength (MPa)		
(a) 72 ± 1 h (b) 168 ± 2 h (c) 672 ± 4 h	40.6 53.7 64.5	Minimum 27 Minimum 37 Minimum 53

3.2 Mix Proportions

In this experimental investigation on standard concrete of M30 grade, cement is replaced with different quantities, 30, 40 and 50% of fly ash by mass of cement besides one mix without fly ash. Ordinary Portland cement of 53 grade (Basha et al. 2014), fine aggregates of zone-II (IS 3025), coarse aggregate as per IS code (IS: 383-1970) provisions, class F fly ash collected from local thermal power plant N.T.P.C, Visakhapatnam, Andhra Pradesh, are used for the present study. The mix proportions for this study is, cementitious material: fine aggregate: coarse aggregate, 1: 1.39: 3.43 with water/binder of 0.48 arrived after several trial mixes as per guide lines of IS: 10262-2009 and IS: 456-2000.

The ingredients of concrete are weighed as per the above proportions and mixed in drum-type concrete mixer as per the IS code specifications and the specimens of cubes of 10 cm, cylinders of 15 cm diameter, 30 cm long and beams of 10 cm × 10 cm × 50 cm are cast on a table vibrator. The slump value and the compaction factor value are found during the experimental work for each mix. Figure 1a and b shown are the photos while mixing the concrete in drum-type mixer and the prism specimens after cast, respectively. The specimens are demoulded after 24 h past cast and kept in a tank for curing for the specified age of 7, 28, 56 and 91 days.

3.3 Testing of the Specimens

The specimens after the prescribed age of curing are tested for compression as per IS 516-specifications in 3000 kN compression testing machine and for flexural strength in 10MT tension testing machine. Specimens are tested for split tension as per IS

(a) (b)

Fig. 1 **a** Mixing of concrete in drum mixer. **b** Specimens of prisms after cast

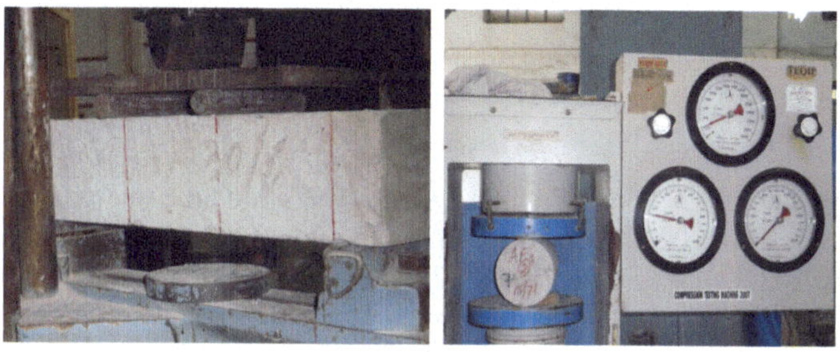

Fig. 2 Testing of prism specimen and cylinder specimen

5816 specifications in compression testing machine. The specimens are also tested for stress–strain in the compression testing machine by using the elastometer. For each test, three specimens are tested, and the average value is considered for the presentation of results. Photos of cylinder specimens during test in split tension and prism specimens for flexure test are shown in Fig. 2.

4 Results and Discussions

The experimental values for compression, split tension and flexural strength are presented below under separate headings.

4.1 Effect of Fly Ash on Compressive Strength

Figure 3 shows the deviation of percentage of residual compressive strength of standard concrete (SC) with percentage of fly ash. From Fig. 3, the values of percentage compressive strength of SC for 0, 30, 40 and 50% fly ash at room temperature, at an

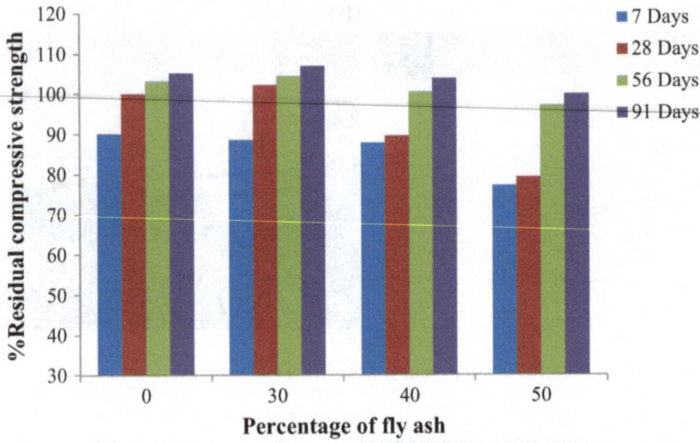

Fig. 3 Variation of percentage of residual compressive strength of standard concrete (SC) with percentage of fly ash

age of 7 days are 90.2, 88.6, 87.9 and 77.2. At the same time, the values at an age of 28 days are 100, 102.2, 89.6 and 79.3. At the same time, the values at an age of 56 days are 103.3, 104.6, 100.6 and 97.2; whereas, the values at an age of 91 days are 105.2, 107, 103.9 and 99.6, respectively.

The compressive strength is increased with increase in fly ash content, at 30% replacement of cement with fly ash; at 40 and 50% replacements the compressive strength values decreased. Similar tendency is observed by Khan et al. (2013).

4.2 *Effect of Fly Ash on Split Tensile Strength*

Figure 4 shows the deviation of percentage of residual split tensile strength of standard concrete (SC) with percentage of fly ash.

From Fig. 4, the values of percentage split tensile strength of SC for 0, 30, 40 and 50% fly ash at room temperature at an age of 7 days are 88.7, 87.7, 84.0 and 77.2. At the same time, the values at an age of 28 days are 100, 102.4, 96.5, 85.9, while the values at an age of 56 days are 102, 103, 100.3 and 92.3; whereas, the values at an age of 91 days are 106.8, 108.6, 104.7 and 101.89, respectively.

The experimental values have revealed that the values of percentage residual split tensile strength enhanced by replacing cement by fly ash of 30% compared to the results of conventional concrete. The similar tendency is followed in the investigations of Khan and Prasad (2015).

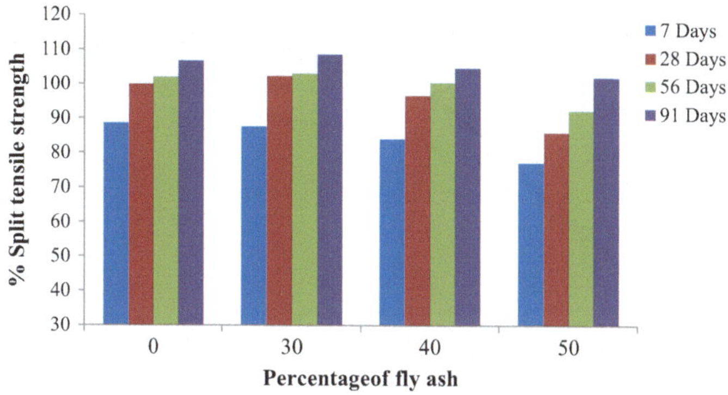

Fig. 4 Deviation of percentage of residual split tensile strength of standard concrete (SC) with percentage of fly ash

4.3 *Effect of Fly Ash on Flexural Strength*

Figure 5 shows the deviation of percentage of residual flexural strength of standard concrete (SC) with percentage of fly ash.

From 5, the values of percentage flexural strength of SC for 0, 30, 40 and 50% fly ash at room temperature at an age of 7 days are 87.6, 82.5, 78.7 and 74.6. At the same time, the values at an age of 28 days are 100, 101, 94.3 and 92.4, while the related values at an age of 56 days are 102.3, 103, 96.5 and 94.4; whereas, the values at an age of 91 days are 104.2, 105, 97.9 and 95.8, respectively.

The test results on the mechanical properties follow a similar trend by earlier investigators. The test results revealed that there is a significant improvement in cube strength in compression, concrete strength in split tension and beam flexural strength by replacing cement with fly ash 30% compared to those of controlled concrete (without fly ash). The percentage residual compressive strength, percentage residual split

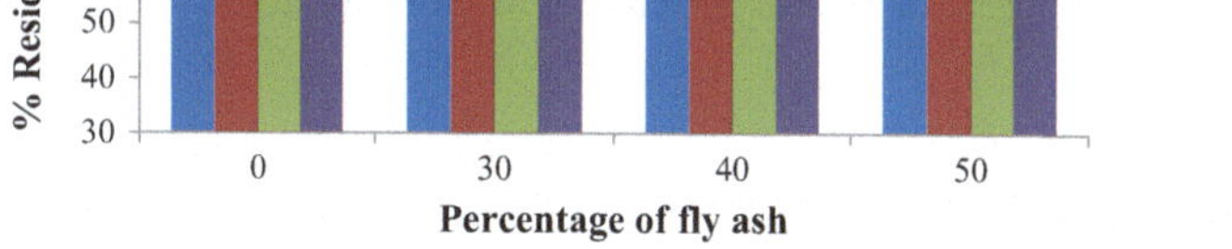

Fig. 5 Variation of percentage residual flexural strength with percentage of fly ash for M30 grade concrete

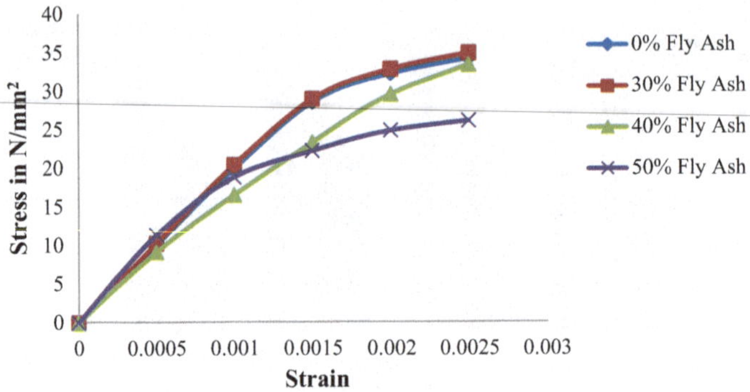

Fig. 6 Variation of stress–strain for different percentages of fly ash for M30 grade concrete

tensile strength and percentage residual flexural strength values decreased compared with those of control concrete and with 40 and 50% replacement of cement with fly ash.

4.4 Stress–Strain Behaviour

The stress–strain behaviour of standard concrete with various replacements of cement 0, 30, 40 and 50% of fly ash is shown in Fig. 6.

The stress–strain response of controlled and fly ash concrete varies linearly and then followed by a parabolic curve till peak value and then decreased prior to failure. The stress–strain curve for SC has fewer slopes compared to HSC which indicates at the same stress, SC has more strain compared to HSC. The stress–strain behaviour of different concretes at room temperature changes widely and mainly depends on water–binder ratio, concrete age during test, the method of curing, the type and amount of aggregates.

The experimental results revealed that the replacing cement with fly ash between 30 and 50%, compared with values for control concrete, the values for concrete containing 30% fly ash have more peak stress value, ductility decreased and stiffness was decreased. For 40 and 50% fly ash replacements, the strengths are decreased compared to controlled concrete and concrete with 30% fly ash. Hence as the percentage of fly ash increased to 40%, ductility and stiffness increased compared to either controlled concrete or fly ash concrete with 30% fly ash.

5 Conclusions

The following conclusions are drawn from the experimental values:

1. Cube strength, cylindrical split tensile strength and beam flexural strength at an age of 7 days are less than those of conventional concrete, due to the delay of hydration process for the specimens with fly ash.
2. The compressive strength improved significantly for the concrete specimens for replacing of cement by fly ash of 30% compared to conventional concrete specimens, while with the 40 and 50% fly ash replacement the values decreased compared to those of conventional concrete.
3. The values of strength of concrete in split tension, cube compressive strength and beam flexural strength of fly ash concrete increase with increase in age of concrete, and the increase is more compared to conventional concrete.
4. The values of strength of concrete in split tension increase for the specimens with 30% replacement of cement by fly ash compared with the values of conventional concrete specimens, while with the 40 and 50% fly ash replacement the values decreased compared to conventional concrete specimens.
5. The flexural strength values improved significantly for the specimens with 30% replacement of cement by fly ash compared to those of conventional concrete specimens, while with the 40 and 50% fly ash replacement the values decreased compared to conventional concrete specimens.
6. The values of beam flexural strength for fly ash concrete increase with increase in age of concrete, and the increase is more compared to conventional concrete.
7. The workability improved with increase in fly ash content, and the slump value increases with increase in fly ash content.
8. The experimental results on stress–strain behaviour revealed that the replacing cement with fly ash between 30 and 50%, compared with those of control concrete, the concrete with 30% fly ash has more peak stress value, ductility decreased and stiffness was decreased. For 40 and 50% fly ash replacements, the strength values are decreased compared with controlled concrete and concrete with 30% replacement of cement with fly ash.
9. Finally, it is concluded that replacement of cement with 30% fly ash in regular works can be recommended and where strength is of minor importance cement can be replaced with either 40% fly ash or 50% fly ash based on the strength requirements.

References

Basha, S. A., Pavithra, P., & Sudarshan reddy, B. (2014). Compressive strength of fly ash based cement concrete. *International Journal of Innovations in Engineering and Technology, 4*(4), 141–156.

Hand Book on *High volume fly ash concrete technology* (3rd ed.) (2005).

IS: 10262-2009, Recommend guidelines for concrete mix design. New Delhi: Bureau of Indian Standards.

IS: 12269, Specifications for 53 grade Portland cement. New Delhi, India: Bureau of Indian Standards.

IS: 383-1970, Specifications for coarse and fine aggregates from natural sources for concrete. New Delhi, India: Bureau of Indian Standards.
IS: 5816-1999, Method of test for splitting tensile strength of concrete. New Delhi, India: Bureau of Indian Standards.
IS: 456-2000, Indian code of practice for plain and reinforced concrete for general building construction. New Delhi: Bureau of Indian Standards,.
IS 3025, Part II Methods of sampling and test for water and waste water for pH.
IS 516:1959 (Reaffirmed 1999), Indian standard methods of tests for strength of concrete. New Delhi. India: Bureau of Indian Standards.
Khan, M. S., & Prasad, J. (2015). Fly ash concrete subjected to thermal cyclic loads. *Fatigue & Fracture of Engineering Materials & Structures, 33*(5), 276–283.
Khan, M. S., Prasad, J., & Abbas, H. (2013). Effect of high temperature on high volume fly ash concrete. *Arabian Journal for Science and Engineering, 38,* 1369–1378.
Malhotra, V. M., & Mehta, P. K. (2005). *High performance high volume fly ash concrete*. Supplementary Cementer Materials for Sustainable Development Inc., Ottawa, Canada.
Okoye, F. N., & Singh, N. B. (2016). Mechanical properties of high volume fly ash concrete. *IOSR Journal of Mechanical and Civil Engineering (IOSR-JMCE), 13*(3), 01–08. Ver. IV (May–June), e-ISSN: 2278-1684, p-ISSN: 2320-334X.
Pandhye, R. D., Doe, N. S. (2016). Cement replace by fly ash in concrete. *International Journals, 5*(Issue special 1), 60–62. ISSN: 2319-6890 (online), 2347-5013, 8&9.
Siddique, R., & Khatib, J. M. (2010). Abrasion resistance and mechanical properties of high volume fly ash concrete. *Materials and Structures, 43*(5), 709–718 (RILEM 2009).

Potential of Silica Sources Including Fly Ash as Green Technology Inputs to Induce Resistance to Biotic and Abiotic Stresses in Crop Plants: Overview

S. Sithanantham, M. Prabakaran and P. Narayanasamy

Abstract Naturally occurring silica source materials, including fly ash, when developed as suitable silica source formulations, can be utilised as green technologies for fortifying the target crop cultivars with induced resistance to biotic and abiotic stresses. Silica sources/products availed from natural or biological sources may have added advantage in organic-certified crop production systems. The potential for utility of applied silica for induced resistance to biotic stresses appears more promising in monocot crops like rice and sugarcane, since they are silica-hungry. There is adequate evidence of silica content in target tissues being associated with genetic variation for host plant resistance to pests in rice and sugarcane. Since farmers tend to select and adopt crop varieties mostly for agronomic attributes like higher yield and crop duration, there is scope for fortifying them with such applied silica by minimising the losses caused by biotic stresses like pests and diseases among such agronomic ally promising and locally popular crop varieties. The major mechanisms by which resistance to biotic stresses like insect pests and diseases could be induced by applied silica include anatomical changes like enhanced silica deposition in the target tissues in epidermal layer as improved structural barrier for their entry and/or biochemical changes due to modifications in metabolic functions including the jasmine acid pathway. Adequate knowledge base is available on the scope for such applied silica products in also countering abiotic stresses such as drought and lodging tendency. There is good scope for locally optimising the application regime of such silica products so to maximise the cost-effectiveness. The results of recent collaborative R&D in sugarcane against borer pests in India are illustrated as case study. The scope for public–private R&D partnership targeting the formulation and market availability of such silica source products as potential components of evergreen revolution in India is indicated. Multidisciplinary and inter-institutional R&D collaboration is reckoned as the promising strategy.

S. Sithanantham (✉) · M. Prabakaran
Sun Agro Biotech Research Centre (SABRC), 3/1798, Main Road, Madanandapuram, Chennai, Tamil Nadu 600 125, India
e-mail: sabrcchennai@yahoo.co.in

P. Narayanasamy
Entomology Department, Faculty of Agriculture, Annamalai University, Chidambaram, Tamil Nadu 608 002, India

S. K. Ghosh and V. Kumar (eds.), *Circular Economy and Fly Ash Management*,
https://doi.org/10.1007/978-981-15-0014-5_10

Keywords Silica sources · Fly ash · Inducing resistance · Biotic stresses · Sugarcane borers

1 Introduction

The International Plant Nutrition Institute (IPNI 2015) has recently listed silicon as a "beneficial substance", which reflects the numerous studies which have established its importance in crop productivity and also in enhancing the resistance of crop plants to biotic and abiotic stresses in crop plants as pointed out by Ma (2004).

2 Role of Silica in Resistance to Insect Pests

While the utilisation of insect resistance identified/developed among genotypes within a crop is reckoned as more desirable and sustainable, there is also good scope for **induced resistance** by applied nutrients as means of fortifying promising crop genotypes through manipulating the plant physiological/biochemical functions manifest as adversity to the development or survival of pest insects (Painter 1968). Reynolds et al. (2017) commented that the discovery that Si plays a role in the resistance of plants to insect herbivores dates back to the early 1950s, yet we are still to fully understand the mechanisms involved. The contributions of applied silica to the physiological processes in inducing resistance to insect herbivores in crop plants have been documented recently by Liang et al. (2015).

The role of silica in induced resistance in crop plants to insect pests has been reviewed by Vasanthi et al. (2014) and the list of insect pests studied in crop plants is furnished in Table 1.

2.1 Role of Silica Against Sugarcane Pests

2.1.1 Studies in Africa

While sugarcane is known to accumulate silica, the applied silica has been found to induce resistance to the sugarcane stalk borer, *Eldana saccharina*, in Africa (Reynolds et al. 2009). The major mechanisms whereby Si could mediate such plant defence against insect herbivores include increased physical (passive) resistance leading to their reduced digestibility and/or increased hardness and abrasiveness (Massey and Hartley 2009), active priming of plant chemical defences by soluble Si, and its interaction with the Jasmonic Acid (JA) signalling pathway, facilitating enhanced production of defensive enzymes (Ye et al. 2013). Vilela et al. (2014) found similar effect of silica in inducing resistance to the sugarcane borer, *Diatraea saccharalis*.

Table 1 Reduction of pest incidence in crops due to silica content

Rice	Stem maggot: *Chlorops oryzae* Green leafhopper: *Nephotettix cinticeps* Brown planthopper: *Nilaparvata lugens* White-backed planthopper: *Sogatella furcifera* Stem borer: *Chilo suppressalis* African striped borer: *Chilo zacconius* Yellow stem borer: *Scirpophaga incertulas* Leaf folder: *Cnaphalocracis medinalis* Gall midge: *Oresolia oryza*
Maize	Stalk borer: *Chilo partellus* Borer: *Sesamia calamistis* Leaf aphid: *Rhapalosiphum maidis* European corn borer: *Ostrinia nutitali* Army worm: *Spodoptera frugiperda*
Muskmelon	Fruit fly: *Bactrocera cucurbitae*
Sorghum	Green bug: *Schizaphis graminum*
Sugarcane	Shoot borer: *Chilo infuscatellus* Stem borer: *Diatraea saccharalis* African stalk borer: *Eldana saccharina*
Wheat	Green bug: *Schizaphis graminum* Hessian fly: *Phytophaga destructor* Whitefly: *Bemisia, tabaci*
Zinnia	Aphids: *Myzus persicae*

Source Vasanthi et al. (2014)

Further, the resistance against *E. saccharina* differed in extent to the thrips, *Fulmekiola serrata* (Keeping et al. 2010).

2.1.2 Case Study of Recent Studies on Sugarcane Borers in India

(i) **Silica association in varietal resistance**: Earlier studies in South India by Rao (1967) and Bhavani et al. (2011) have shown that varietal resistance to the early shoot borer (ESB), *Chilo infuscatellus* had positive relationship with leaf silica levels among sugarcane varieties.

(ii) **Applied silica effects on borer tolerance**: Recent exploratory studies have shown the potential for applied potassium silicate in significantly reducing the infestation by the ESB (*C. infuscatellus*) which was also reflected in higher leaf silica content (Prabakaran et al. 2015). They found that when applied as potassium silicate or sodium silicate, there was significant reduction in ESB infestation, which was next only to that of chemical insecticide (monocrotophos), and even superior to sprays of the recommended microbial biopesticide-Bt, whereas the other two sources of potassium (chloride or sulphate) could only moderately reduce the ESB incidence (Fig. 1).

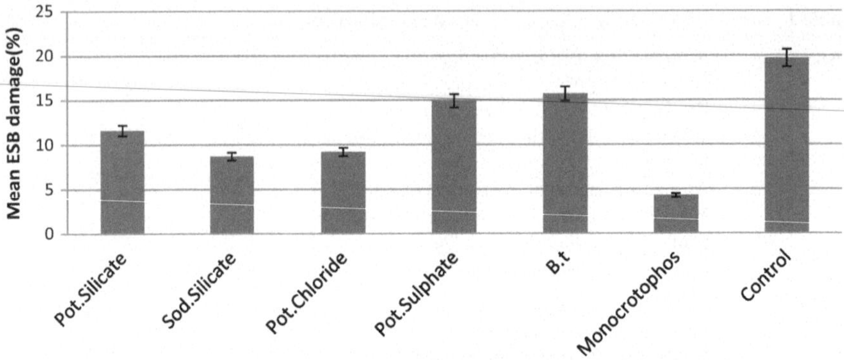

Fig. 1 Relative impact of potassium silicate on sugarcane borer-ESB, 2015. *Source* Prabakaran et al. (2015)

Prabakaran et al. (2017) further compared three silica sources applied as foliar spray or soil drench during tillering phase and found that the cumulative percentage of shoot borer damage (dead-hearts) over 6 and 8 week periods differed significantly, with four applications of potassium silicate as spray or drench causing greatest reduction in per cent dead-hearts and also comparable in the borer levels to four sprays of the biopesticide—*Bacillus thuringiensis*. Further, SEM study confirmed the translocation and deposition of additional silica in the leaf epidermis, which could have been the ultra-structural barrier for the freshly hatched larvae to penetrate the epidermal layer.

2.2 Role of Silica in Managing Rice Pests

The incidence of stem maggot, green leafhopper, brown planthopper, white-backed planthopper and leaf folder in rice being reduced due to silica nutrition was reported by Sawant et al. (1994). Arivuselvi (2014) ascertained that the rice plants applied in foliar/soil routes showed significantly reduced the incidence and damage by sucking insects pests viz. brown planthopper, green leafhopper and ear head bug as well as internal feeders viz. stem borer and gall midge and as well as leaf feeder viz. leaf folder and spiny beetle as compared to untreated check. Basal application of calcium silicate with foliar spray of SMS recorded more silica content, polyphenol oxidase and PALase which in turn significantly reduced the incidence of rice stem borer to a tune of 74% (Chandramani et al. 2015, 2018). Tripathy and Rath (2017) conducted field study and laboratory analysis in rice, concluding that orthosilicic acid @ 4 ml/l as foliar spray four times was the best treatment resulting in enhanced the desirable yield attributes of rice plants and produced higher grain yield, besides significant reduction in yellow stem borer infestation. Silicon was found to play a major role in increasing the leaf sheath compactness and cell wall thickening apparently causing

Table 2 Correlation of silica uptake versus plant morphological and stem borer infestation parameters and grain yield in rice

Parameters	Correlation coefficient®
Si content versus plant height	0.70*
Si content versus tiller number	0.77*
Si content versus number of leaves per hill	0.83**
Si content versus stem diameter	−0.81**
Si content versus number of effective tillers per hill	0.80**
Si content versus dead heart (maximum DH stage)	−0.72*
Si content versus white ear head (maximum WEH stage)	−0.69*
Si content versus tunnel length (maximum DH stage)	−0.89**
Si content versus tunnel length (WEH stage)	−0.54
Si content versus grain yield	0.95**

Source Tripathy and Rath (2017)
*Significant at 5% leve, **Significant at both 1 and 5% level

impaired larval penetration, thereby reducing stem borer infestation (Table 2). While orthosilicic acid was adjudged as the most effective silica product, it was followed in rank by calcium silicate, steel slag and fly ash.

2.3 Tritropihic Interactions—Role of Silica

Silicon influence on the tritrophic interactions in wheat plants involving the green bug, *Schizaphis graminum* and its natural enemies, *Chrysoperla externa* and *Aphidius colemani* has been elucidated. There was information on indirect plant defence based on augmented release of herbivore-induced plant volatiles (HIPVs) that also attract natural enemies of the herbivore (Kvedaras et al. 2010).

2.4 Effects on Stored Products Insect Pests

Silica formulations have also been known to play a role in the management of stored products insect pests (Vasanthi et al. 2014) and the available information is summarised in Table 3.

Table 3 Pests of stored products controlled by silica formulations

Rice husk ash	*Callosobruchus analis*
Wood ash	*Callosobruchus analis, C. chinensis, C. maculatus*
Silica	*Tribolium confusum, Sitophilus granarius, Rhyzopertha dominica, Tenebrio molitor, Plodia interpunctella, Coleomegilla maculata, Leptinotarsa decemlineata*
Silica nanoparticles	*Sitophilus oryzae*
Silica aerogel	*Prostephanus truncatus, Oryzaephilus mercator*

Source Vasanthi et al. (2014)

Table 4 Known suppression of non-insect pests by silica application/levels

Nematodes	
Meloidogyne sp.	Rice
Meloidogyne exigua	Coffee
Meloidogyne incognita	Cucumber
Pratylenchus zeae, Helicotylenchus dihystera	Sugarcane
Mites	
Leaf spider mite *Tetranychus* sp.	Rice
Red spider mite *Tetranychus urticae*	Brinjal

Source Vasanthi et al. (2014)

2.5 *Effects of Silica on Non-insect Pests of Crops*

It has also been known that several non-insect pests are also affected by silica levels in crops (Vasanthi et al. 2014), which are listed in Table 4.

3 Role of Silica in Managing Plant Diseases

Considerable information has also been assembled on role of silica in reducing the incidence levels of crop diseases caused by plant pathogens, and the illustration of such known reports is provided in Table 5.

4 Role of Silica in Abiotic Stresses

The known cases of role of silica in countering abiotic stresses in crop plants are illustrated in Table 6.

Table 5 Silicon induced suppression/resistance against plant diseases

Banana	Panama wilt: *Fusarium oxysporium* f.sp. *cubense* Root rot: *Cylindrocladium spathiphylli*
Carrot	Cercospora blight: *Cercospora carotae* Alternaria blight: *Alternaria dauci*
Coffee	Frog's eye spot: *Cercospora, coffeicola*
Cucumber	Powdery mildew: *Sphaerotheca fuliginea* Crown and root rot: *Pythium ultimum*
French bean	Cow pea rust: *Uromyces phaeseolitypia*
Melon	Fruit decay: *Alternaria alternate, Fusarium semitectum, Trichothecium roseum*
Rice	Rice blast: *Pyricularia oryzae* Grain discoloration: *Bipolaris, Fusarium* Brown spot: *Cochliobolus miyabeanus,* Leaf scald: *Monographella albescens* Sheath blight: *Rhizoctonia solani* Stem rot: *Magnaporthe salvinii* Bacteriallea fb light: *Xanthomonas oryzae* pv. *oryzae*
Soybean	Stem canker: *Diaporthe phaseolorum* F.sp. *meridiunalis* Soybean Sudden death syndrome: *Fusarium solani* Seedling damping-off: *Fusarium semitectum* Downy mildew: *Peronospora manshurica,* Frog's eye spot: *Cercospora sojina*
Sugarcane	Sugarcane rust: *Leptosphaeria Sacchari* Ringspot *Puccinia melanocephala*
Wheat	Powdery mildew *Erysiphe graminis, Oidium monilioides, Blumeria graminis* sp., *tritici* Leaf spot: *Mycosphaerella pinodes*

Source Vasantha et al. (2014)

Table 6 Silicon nutrition and mitigation of different abiotic stress in crops

Physical stress	Lodging, drought, high temperature, freezing, UV radiation, etc.
Chemical stress	
Salinity	Rice, wheat, mesquite, maize
Mn toxicity	Cucumis, bean
Al toxicity	Rice
Cu toxicity	Arabidopsis
Fe toxicity	Sugarcane
Cd toxicity	Rice, maize
Zn toxicity	Maize

Table 7 Silicon Sources including fly ash applied to augment the yield in different crops

Cucumber	Sodium metasilicate
Maize	Coal fly ash
Mustard	Coal fly ash
Potato	Ligno-silicon
Rice	Calcium silicate, Ca–Mg silicate, sodium metasilicate, silicate slag, rice hull, rice hull ash, rice straw, poha industry waste, lignite fly ash, coal fly ash
Sugarcane	Calcium silicate, calcium metasilicate, Ca–Mg silicate, silicate slag, electric furnace slag, steel slag, Tennessee Valley Authority (TVA) slag, bagasse furnace ash, basalt, cement
Soybean	Poha industry waste
Wheat	Silicic acid, coal fly ash, Poha industry waste

Source Vasantha et al. (2014)

5 Fly Ash as Potential Silica Source for Enhancing Crop Yield

The utility of fly ash and other silica sources in improving the crop yields is illustrated in Table 7.

6 Recommendations for Future R&D

Based on the past literature and the on-going R&D towards improved utilisation of silica sources, the following initiatives are recommended:

1. Assessing the scope for alternative silica sources—both organic source (like rice grain husk and different industrial by-products like fly ash) in inducing resistance to major insect pest and disease problems in major crops across major agro-ecological regions in India.
2. Evaluate the scope to optimise silica application regimes based on local soil silica status, so to also capture the benefits in countering abiotic stresses like drought and lodging.
3. To seek the synergistic benefits of the twin goals of safer disposal of silica-rich waste products and fortify our target crops with induced resistance to biotic and abiotic stresses.
4. There is scope for multidisciplinary approach and to evolve public–private partnership models in R&D and further commercialisation leading to win-win scenario.

Acknowledgements The authors are grateful to the conference organizers for considering this paper for inclusion in the proceedings. Special thanks to Dr. Vimal Kumar for his keen support for this effort.

References

Arivuselvi, A. (2014). *Silica induced resistance against major insect pests of rice* (M.Sc. (Ag.) thesis). Tamil Nadu Agricultural University, Madurai, Tamil Nadu, India.

Bhavani, B., Reddy, K. D., Rao, N. V., & Lakshmi, M. B. (2011). Biochemical basis of resistance in sugarcane to early shoot borer, *Chilo infuscatellus*. *Indian Journal of Plant Protection, 39*(4), 264–270.

Chandramani, P., Arivuselvi, A., Balasubramanian, P., & Nalini, R. (2015). Effect of silica induced physiological factors on major pests of rice. In *Proceedings of International Conference on Innovative Insect Management Approaches for Sustainable Agro Ecosystem (IIMASAE)*, AC & RI, Madurai, India (pp. 249–250).

Chandramani, P. (2018). Impact of silicon induced physiological factors on rice stem borer, *Scirpophaga incertulas*. *Indian Journal of Natural Sciences, 9*(51), 15339–15342.

International Plant Nutrition Institute [IPNI]. (2015). *Nutri-Facts. Silicon No. 14*. Available at: http://www.ipni.net/publication/nutrifacts-na.nsf/0/A7B4AB4D35C153BF85257.ECE006E0E34/$FILE/NutriFacts-NA-14.pdf. Accessed April 4, 2016.

Keeping, M. G., Mcfarlane, S. A., Sewpersad, N., & Rutherford, R. S. (2010). Effects of silicon and plant defence inducers on sugarcane yield parameters, *Eldana saccharina* Walker (Lepidoptera: Pyralidae) and thrips, *Fulmekiola serrata* Kobus (Thysanoptera: Thripidae). *Proceedings of the Annual Congress-South African Sugar Technologists' Association, 83*, 271–275.

Kvedaras, O. L., An, M., Choi, Y. S., & Gurr, G. M. (2010). Silicon enhances natural enemy attraction and biological control through induced plant defences. *Bulletin of Entomological Research, 100*(3), 367–371.

Liang, Y., Nikolic, M., Belanger, R., Haijun, G., & Song, A. (2015). *Silicon in agriculture: From theory to practice*. Dordrecht: Springer.

Ma, J. F. (2004). Role of silicon in enhancing the resistance of plants to biotic and abiotic stresses. *Soil Science and Plant Nutrition, 50*, 11–18.

Massey, F. P., & Hartley, S. E. (2009). Physical defences wear you down: Progressive and irreversible impacts of silica on insect herbivores. *Journal of Animal Ecology, 78*, 281–291.

Painter, R. H. (1968). *Insect resistance in crop plants* (p. 520). USA: University Press of Kansas.

Prabakaran, M., Sithanantham, S., Subramanian, K. S., Vijayaprasad, P., Punna Rao, B. V., & Babu, B. (2017). Silica application effect on sugarcane early shoot borer. In *Proceedings of International Symposium on "Sugarcane Research Since Co205:100 Years and beyond" (SucroSym 2017)*, 18th–21st September 2017, Sugarcane Breeding Institute, Coimbatore, Tamil Nadu (pp. 393–395).

Prabakaran, M., Sithanantham, S., Kandasamy, R., Subramanian, K. S., Gunasekaran, K., & Kannan, M. (2015). Foliar application effect of silicon in sugarcane early shoot borer infestation. *Hexapoda, 22*, 59–62.

Rao, S. D. V. (1967). Hardness of sugarcane varieties in relation to shoot borer infestation. *Andhra Agricultural Journal, 14*, 99–105.

Reynolds, O. L., Padula, M. P., Zeng, R., & Gurr, G. M. (2017). Silicon: Potential to promote direct and indirect effects on plant defense against arthropod pests in agriculture. *Frontiers in Plant Science, 7*, 744.

Reynolds, O. L., Keeping, M. G., & Meyer, J. H. (2009). Silicon-augmented resistance of plants to herbivorous insects: A review. *Annals of Applied Biology, 155*, 171–186.

Sawant, A. S., Patil, V. H., & Savant, N. K. (1994). Rice hull ash applied to seedbed reduces dead hearts in transplanted rice. *International Rice Research Notes, 19*(4), 21–22.

Setamou, M., Schulthess, E., Bosque-Perez, N. A., & Odijo-Thomas, A. (1993). Effects of plant nitrogen and silica on bionomics of *Sesamia calamistis* (Lepidoptera: Noctuidae). *Bulletin of Entomological Research, 83*, 405–411.

Sithanantham, S., Kandasamy, R., Prabakaran, M., Judy, S., & Manikandan, K. R. (2015). Multifaceted factory level R&D model to locally maximise the economic benefit of eco-friendly sugarcane borer management methods. *SISSTA Sugar Journal, 45*, 37–49.

Tripathy, S., & Rath, L. K. (2017). Silicon induced resistance expression in rice to Yellow stem borer. *Journal of Entomology and Zoology Studies, 5*(5), 12–15.

Vasanthi, N., Saleena, L. M., & Anthoni Raj, S. (2014). Silicon in crop production and crop protection—A review. *Agricultural Reviews, 35*(1), 14–23.

Vilela, M., Moraes, J. C., Alves, E., Santos-Cividanes, T. M., & Santos, F. A. (2014). Induced resistance to *Diatraea saccharalis* (Lepidoptera: Crambidae) via silicon application in sugarcane. *Revista Colombiana de Entomologia, 40,* 44–48.

Ye, M., Song, Y., Long, J., Wang, R., Baerson, S. R., & Pan, Z. (2013). Priming of jasmonate-mediated anti-herbivore defense responses in rice by silicon. *Proceedings of the National Academy of Sciences of the United States of America, 110,* E3631–E3639.

Fly Ash as a Source of Silicon for Mitigating Biotic Stress and Improving Yield and Changes in Biochemical Constituents and Silicon in Rice Under Abiotic Stress

P. Balasubramaniam

Abstract Rice (*Oryza sativa* L.) is the most irreplaceable staple food for over three billion people of Asia. In Tamil Nadu, it is being grown throughout the year at different seasons under low land submerged conditions. As a result of continuous cropping, removal of silicon from soil leads to its deficiency. The rice crop shows the largest uptake of silicon playing a major role in mitigating the biotic and abiotic stresses. The fly ash is the storage of silicon in significant amount irrespective of different sources of fly ash, viz. fly ash from sugar factory, thermal power station and modern rice mill. The availability of fly ash is plenty, which poses a problem to the environment. Hence, the present study was carried out to explore the possibility of utilizing the fly ash as a source of silicon for mitigating the biotic and abiotic stress in rice. In view of the above, field experiments were conducted in low silicon soils to examine the effect of fly ash @ 25 t ha^{-1} with silicate solubilizing bacteria @ 2 kg ha^{-1} and farm yard manure @ 12.5 t ha^{-1}, with graded level of soil test-based potassium on growth and yield, incidence of pests, silicon uptake and changes in biochemical constituents, viz. total phenol, OD phenol, reducing, non-reducing sugars and proline in different plant parts such as leaf blade, leaf sheath cum stem and in ear head at different growth stages of the rice cultivar BPT 5204 under induced drought and flood conditions. The experiments were conducted at Agricultural Engineering College and Research Institute Farm, Kumulur, Trichy District, during 2012 in split plot design with two replications. The results revealed that application of fly ash with silicate solubilizing bacteria + farm yard manure with 100% soil test-based potassium increased the rice grain yield by 22 and 10.3%, under induced drought and flood stress conditions, respectively, over control. The biochemical constituents played an important role in reducing the pest incidence in induced drought and flood stress. However, the effect was more pronounced under induced drought over flood stress condition. Hence, the application of fly ash with silicate solubilizing bacteria + farm yard manure with 100% soil test-based potassium helps to improve the growth

P. Balasubramaniam (✉)
Department of Soil Science and Agricultural Chemistry, Anbil Dharmalingam Agricultural College and Research Institute, Tamil Nadu Agricultural University, Tiruchirappalli 620 027, Tamil Nadu, India
e-mail: balu_tnau@yahoo.co.in

S. K. Ghosh and V. Kumar (eds.), *Circular Economy and Fly Ash Management*,
https://doi.org/10.1007/978-981-15-0014-5_11

and yield of rice. It also helps in mitigating the biotic stresses in rice by the influence of Si under induced abiotic stress, viz. drought and flood. Other biochemical constituents were also favorably correlated with the reduced incidence of pests and increased yield of both grain and straw.

Keywords Fly ash · Biotic and abiotic stress · Silicon · Soil test-based potassium · Rice

1 Introduction

Rice is the indispensable food for over three billion people of Asia. At a hastening growth rate of 1.8% of the population in India, if rice requirement is to cope up with the population, the yield level of rice has to be triggered by 25–30% from the present level of 1.9 t ha^{-1} if the country is to remain self-sufficient. In Tamil Nadu, rice is grown in an area of 20.16 L ha with the production of 62.53 L Mt and with the average productivity of 3102 kg ha^{-1}. Rice is a silicicolous plant that absorbs Si in the form of monosilicic acid (H_4SiO_4) through active aerobic respiration. Silicon is the second most inexhaustible element in the earth's crust with soils encompassing almost 32% Si by weight (Lindsay 1979). Silicon deposited in the walls of epidermal cells after absorption by plant contributes considerably to the strength. The epidermal cell walls become effective barriers against both fungal infections and water loss by cuticular transpiration when impregnated with a firm Si layer (Jones and Handreck 1967). Djamin and Pathak (1967) have found that the incisor region of the mandibles of stem borer larvae fed on rice plants with high Si content was more damaged. Maxwell et al. (1972) and Panda et al. (1977) found that the infestations of rice stem borer were markedly reduced by adding silicon to the soil. Tayabi and Azizi (1984) concluded that the application of silica at 1 t ha^{-1} reduced the population density of stem borer. Development of resistance in insects becomes a major problem due to the indiscriminate use of pesticides. Silicon is considered most important nutrient elements in conferring resistance to biotic stresses, viz. insect pests, nematodes and diseases and abiotic stresses, viz. drought, lodging, salinity, waterlogging and nutrient imbalances in the soil. Rice is known as Si accumulator, and the plants benefit from Si nutrition thereby mitigate biotic (pests and diseases) and abiotic stress (drought and flood) in rice. It is estimated that the rice crop producing a grain yield of 5 t ha^{-1} will normally remove from the soil of 230 to 470 kg Si ha^{-1}. Salim and Saxena (1992) found that at higher levels of silicon, metamorphism of plant hopper to nymphs to adult was reduced and there was a decrease in adult longevity and female fecundity. It has been reported that silicon controls insect pests such as the stem borer, brown planthopper, green leafhopper, whitebacked plant-hopper and non-insect pests, for instance, spider mites (Ma and Takahashi 2002). Although fly ash (FA) contains 0.2–0.3% potassium and 15–60% SiO_2, their availability to rice crop needs to be explored in detail. Research so far carried out did not make an investigation on inducing resistance to biotic and abiotic stress relevant to Si. Not much work has been carried out in integrating the FA with silicate solubilizing bacteria (SSB) and farm yard manure (FYM) with graded levels of soil test-based potassium (STBK) in rice. Hence, the present investigation has been carried out with the objectives of studying the effect of FA, SSB and FYM with graded levels of STBK on growth and yield of rice, incidence of major pests' changes in Si and other biochemical constituents in rice under induced drought and flood stresses.

2 Materials and Methods

Field experiments were done with rice variety BPT 5204 under induced drought and flood stress for 20 days during the tillering stage. The experiments were conducted in a split plot design with two replications. The main plot treatments include M1—control, M2—fly ash (FA) @ 25 t ha^{-1} + silicate solubilizing bacteria (SSB) @ 2 kg ha^{-1}, M3—FA @ 25 t ha^{-1} + farm yard manure (FYM) @ 12.5 t ha^{-1}, M4—FA @ 25 t ha^{-1} + SSB @ 2 kg ha^{-1} + FYM @ 12.5 t ha^{-1} which were followed, and subplot treatments were graded level, viz. 0, 25, 50, 75 and 100% of soil test-based potassium (STBK). The FA was applied before transplanting followed by inoculation of SSB and incorporation FYM. The initial experimental field soils were collected and analyzed for various physicochemical properties by using standard procedure; similarly, the FA used for the experiment was also analyzed for its characterization (Jackson 1973), and the results are given in Table 1. The plant samples, viz. leaf, leaf sheath cum stem and ear head during different growth stages, viz. flowering and maturity stage were collected. The fresh plant samples were used for the analysis of biochemical constituents, viz. total and *ortho*-dihyric phenols, reducing and non-reducing sugars and proline by using standard procedures (Mahadevan and Sridhar 1986). The oven-dried plant samples, grain and straw at 70 °C were used for estimation of silicon. The plant sample of 0.1 g was digested in a mixture of 7 ml of HNO_3 (62%), 2 ml of hydrogen peroxide (H_2O_2) (30%) and 1 ml of hydrofluoric acid (46%) kept in for 10–15 min for predigestion. The samples were digested using microwave digester (Microwave reaction system Antonpaar Multiwave 3000 solv) with following program 500 W for 17 min with a ramp at 10 °C per minute to reach the temperature of 150 °C, 500 W for 10 min for holding the temperature of 150 °C and venting for 10 min. The digested samples were diluted to 50 ml with 4% boric acid (Ma et al. 2001). The Si concentration in the digested solution was determined by transferring 0.1 ml of digested aliquot to a plastic centrifuge tube, added with 3.75 ml of 0.2 N HCl, 0.5 ml of 10% ammonium molybdate and 0.5 ml of 20% tartaric acid and 0.5 ml of reducing agent 1-amino-2-napthol-4-sulfonic acid (ANSA), and the volume was made up to 12.5 ml with distilled water and kept it for one hour. After one hour, the absorbance was measured at 600 nm with a UV–visible spectrophotometer. Similarly, standards (0, 0.2, 0.4, 0.8 and 1.2 ppm) were prepared by using 1000 ppm of silicon stock standard obtained from Merk by following the same procedure. The observations on pest incidence/population were observed under natural condition. The incidence of major pests, viz. ear head bug which was observed as number per

Table 1 Initial characterization of experimental soil and fly ash

S. No.	Particulars	Drought stress field experimental soil (AEC&RI Field No. C2)	Flood stress field experimental soil (AEC&RI Field No. B4)	Fly ash
A.	*Physical properties*			
1	Bulk density ($Mg\ m^{-3}$)	1.5	1.3	1.27
2	Particle density ($Mg\ m^{-3}$)	2.2	2.5	1.99
3	Total porosity (%)	35.4	32.1	42
4	Maximum water holding capacity (%)	29.1	28.7	33
5	Mechanical composition Sand (%) Silt (%) Clay (%) Texture	62.5 24.1 9.2 SL	63.2 23.5 8.8 SL	24.15 62.25 6.25 Si L
B	*Chemical properties*			
1	$pH_{1:2.5}$ pH_s	7.7	7.2	9.1 8.1
2	$EC_{1:2.5}$ (dSm^{-1}) EC_e	0.12	0.49	0.5 0.24
3	Cation exchange capacity (c mole(p^+)kg^{-1})	24	21.8	2.1
4	Organic carbon ($g\ kg^{-1}$)	4.3	5.4	1.1
5	Available nitrogen (Alkaline permanganate N) ($kg\ ha^{-1}$)	106.6	94	–
6	Available phosphorus (Olsen's P) ($kg\ ha^{-1}$)	22	22	–
7	Available potassium (NH_4OAc K) ($kg\ ha^{-1}$)	240	155	36.5
8	Available silicon (NaOAc pH4.0 Si) ($mg\ kg^{-1}$)	40	38	215

ear head and stem borer was observed as a percentage of dead heart before flowering and percentage of the white ear after flowering as below:

$$\text{Stem borer damage as dead heart/white ear (\%)} = \frac{\text{Number of affected tillers/meter}^2}{\text{Total number of tillers/meter}^2} \times 100$$

The yield of grain and straw of both the drought and flood stress experiments was assessed at harvest and expressed at 14% moisture.

3 Results and Discussion

3.1 Effect of FA on the Yield of Rice Under Drought and Flood Stress (Figs. 1 and 2)

The results revealed that the highest grain yield of 6017 kg ha^{-1} was recorded by the addition of FA + SSB + FYM with 100% STBK under drought stress and 6178 kg ha^{-1} under flood stress condition. Application of FA with SSB and FYM with 100% potassium showed a straw yield of 7632 kg ha^{-1} under drought stress and 7428 kg ha^{-1} under flood stress condition. The yield increase of 22% under drought and 10.3% under flood stress over control was observed. The increase in yield might be due to positive relation of yield contributing factors, viz. thousand grain weight, number of filled grains per panicle, number of productive tillers per hill by the addition of FA + SSB + FYM. The graded levels of 100% STBK recorded the highest grain yield. Under the drought stress, the yield reduction was nullified by the supply of Si and K through the addition of FA + SSB + FYM with 100% K. This might be due to the fact that under induced drought stress, the reactive oxygen species (ROS) production enhances the K demand to maintain photosynthesis and protect chloroplast from oxidative damage (Mengel and Kirkby 2001). E1-Hadi et al. (1997) concluded that improvement of Si and K nutritional status of plants seems to be of great importance for sustaining high yields, and they revealed that the decrease in grain yield from restricted irrigation could be greatly eliminated by increase in Si and K supply for plants. The highest increase of yield under drought was attributed to the contribution of FA + SSB + FYM with 100% K toward mitigating the stress effect in the treated plot where in the control, it could not have contributed, hence, more yield reduction in control was observed. This was not happened under induced flood stress as rice grown under flooded condition, just stagnation of water to the land submergence alone, will not decrease the yield level. This might be due to the effective utilization of Si and K released from the applied FA with SSB and FYM with graded level of STBK. The results were corroborated with the findings of Mallika et al. (2000) Sarwar et al. (2008), Chandramani et al. (2009) and Arivazhagan et al. (2011).

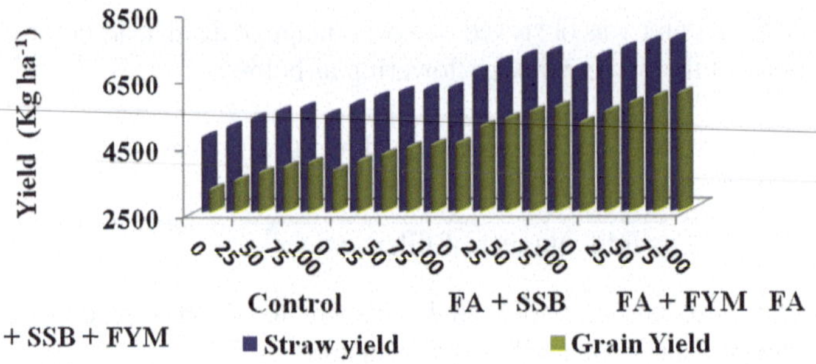

Fig. 1 Effect of FA with SSB + FYM with graded levels of STBK on grain and straw yield in rice under induced drought stress

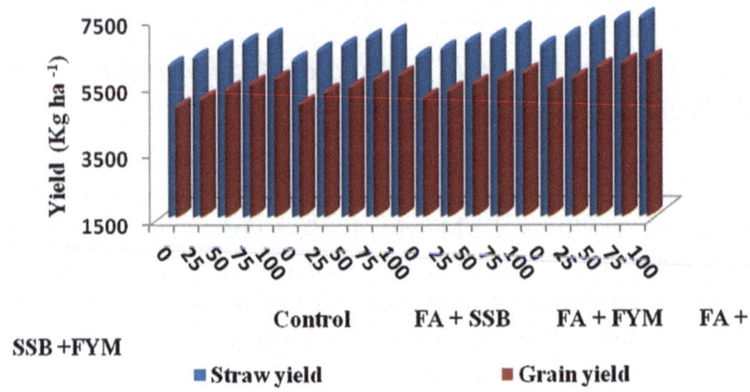

Fig. 2 Effect of FA with SSB + FYM with graded levels of STBK on grain yield and straw yield in rice under induced flood stress

3.2 *Effect of FA on Si Uptake of Rice Under Drought and Flood Stress (Figs. 3 and 4)*

Under drought stress condition, the uptake of Si in rice plants was highly significant at the harvesting stage, and it varied from 161.49 to 296.94 kg ha^{-1} in straw. The highest mean straw uptake (281.74 kg ha^{-1}) was observed due to the application of FA with SSB and FYM followed by FA with FYM (257.80 kg ha^{-1}), FA with SSB (196.11 kg ha^{-1}) and control 178.13 kg ha^{-1}. Among the graded levels of STBK, the interaction of different treatments did not exhibit a significant difference in the Si uptake. The Si uptake in the grain was significantly varied from 20.45 to 75.52 kg ha^{-1}. The highest mean Si uptake of 70.25 kg ha^{-1} in the grain was recorded by the application of FA with SSB and FYM followed by FA with FYM (55.35 kg ha^{-1}), FA with SSB (39.88 kg ha^{-1}) and control (24.50 kg ha^{-1}). The uptake of Si was positively and significantly correlated with grain yield. Under flood stress condition, the highest mean straw uptake (266.88 kg ha^{-1}) was observed due to the application of FA with SSB and FYM followed by FA with FYM (247.55 kg ha^{-1}).

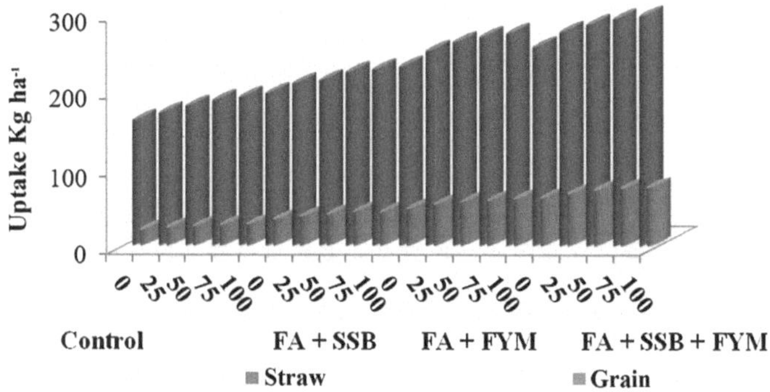

Fig. 3 Effect of FA with SSB + FYM with graded levels of STBK on Si uptake in straw and grain in rice under induced drought stress

The control recorded the lowest straw uptake of 214.45 kg ha^{-1}. Among the graded levels of STBK, the highest mean Si uptake of 256.77 kg ha^{-1} was registered by the application of 100% STBK compared to the control which recorded the lowest uptake of 225.18 kg ha^{-1}. The Si uptake in grain was significantly varied from 29.72 to 69.49 kg ha^{-1}. The highest mean uptake of 64.71 kg ha^{-1} in the grain was recorded by the application of FA with SSB and FYM followed by FA with FYM (58.86 kg ha^{-1}). The control recorded the least uptake of 34.09 kg ha^{-1} in rice grain. Among the graded levels of STBK, application of 100% STBK recorded the highest mean uptake of 51.74 kg ha^{-1} over the rest of the treatments. The uptake of Si in rice was positively correlated with grain yield. Among the different treatments, FA with SSB and FYM recorded the highest Si uptake in both grain and straw under induced drought as well as flood stress condition. There was a significant increase of Si uptake by the application of graded level of STBK. The increase in Si uptake by the addition of 100% STBK might be due to the fact that potassium deficiency reduces the accumulation of Si in the epidermal cells of the leaf blades. The results were confirmed with the findings of Balasubramaniam (2003), Hong et al. (2006), Lee et al. (2006) and Chandramani (2010).

3.3 *Effect of FA on Incidence of Pests Under Drought and Flood Stress*

1. **Stem borer—(*Scirpophaga incertulas* Walker)** (Table 2).

The stem borer incidence was observed at the milky stage and recorded the incidence of stem borer as white ear. The addition of FA + SSB + FYM with 100% STBK recorded the lowest of 0.4% white ear over control which recorded the highest percent (7.5%) which clearly indicated the incidence of white ear was reduced by 46.7% under drought stress. Similar results were also obtained under flood stress condition

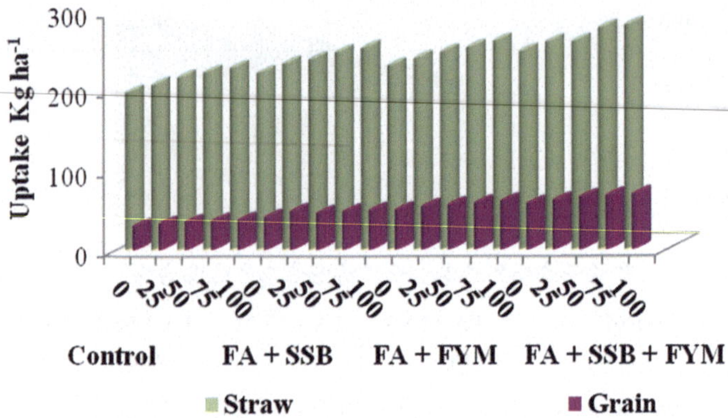

Fig. 4 Effect of FA with SSB + FYM with graded levels of STBK on Si uptake in straw and grain in rice under induced flood stress

Table 2 Effect of FA with SSB + FYM with soil test-based K on incidence of stem borer and ear head bug at milky stage of rice under induced drought and flood stress

Treatments	Drought (milky stage)		Flood (milky stage)	
	Stem borer (white ear %)	Ear head bug (Number per ear head)	Stem borer (white ear %)	Ear head bug (Number per ear head)
Control	7.5 (2.73)	0.5 (0.70)	7.55 (2.74)	0.6 (0.77)
FA @ 25 t ha^{-1} + SSB + FYM +0% STBK	2.26 (1.50)	0.2 (0.54)	2.32 (1.52)	0.50 (0.70)
FA @ 25 t ha^{-1} + SSB + FYM +25% STBK	1.64 (1.28)	0.2 (0.43)	1.63 (1.27)	0.15 (0.38)
FA @ 25 t ha^{-1} + SSB + FYM +50% STBK	1.10 (1.04)	0.15 (0.38)	1.1 (1.05)	0.1 (0.32)
FA @ 25 t ha^{-1} + SSB + FYM +75% STBK	1.20 (1.09)	0	1.19 (1.09)	0
FA @ 25 t ha^{-1} + SSB + FYM + 100% STBK	0.40 (0.62)	0	0.45 (0.66)	0
SED	0.13	0.05	0.13	0.21
CD ($P = 0.05$)	0.29	0.10	0.28	NS

with respect to incidence of stem borer. The stem borer incidence was negatively and significantly correlated with the Si ($r = -0.54$) and K ($r = -0.81$) content in leaf sheath cum stem. This might be due to the application of FA that enhances Si and K content results in the damage of mandibles of larvae of the rice stem borer (Djamin and Pathak 1967). The deposition of Si and K at flowering stage might also contribute for the reduced incidence of stem borer. Narayanaswamy (1994) reported that the main cause for the death of insects due to FA application was wearing of mandibles and main feeding organs of insects which resulted in functionless mandibles so that insect like leaf folder, stem borer die without food. Further increased phenolics and biophysical factors, viz. epicuticular wax and leaf sheath thickness also have contributed to a reduction in the stem borer incidence. Similar results were also corroborated with the findings of Balasubramaniam (2003) and Chandramani et al. (2009).

2. **Ear head bug (*Leptocorisa acuta* Thumb)** (Table 2).

The application of FA + SSB + FYM showed significant reduction in ear head bug population during the milky stage and recorded no ear head bug population in the plots receiving FA + SSB + FYM with 100% STBK; whereas, the control recorded the highest ear head bug population of 0.5/5 hill. A similar observation was also made under flood stress. The less ear head bug by the addition of FA + SSB + FYM with 100% STBK might be due to the Si present in the FA which gets deposited in the plant which in turn inhibits the feeding activity of ear head bug. The antibiosis mechanism of resistance in rice to ear head bug might be due to the presence of defensive chemicals like phenol. Further, the increased content of Si, K and phenols in the ear head at the flowering stage by the addition of FA + SSB + FYM with STBK might also be contributed for anti-feeding action of ear head bug at flowering stage. The ear head bug population was negatively correlated with Si ($r = -0.7$), K ($r = -0.9$) and total phenol ($r = -0.94$) at flowering stage. The results were coping up with the findings of Meyer and keeping (2005).

3.4 *Effect of FA on Changes in Biochemical Constituents Under Drought and Flood Stress*

1. **Total phenol** (Tables 3 and 4)

The total phenol content in different plant parts showed a significant difference between the treatments at flowering stage. The highest total phenol content of 2.25 mg g^{-1} in leaf blade, 0.91 mg g^{-1} in leaf sheath cum stem and 1.04 mg g^{-1} in ear head was recorded due to the addition of FA + SSB + FYM with 100% STBK; whereas, the control recorded 1.8 mg g^{-1} in leaf blade, 0.4 mg g^{-1} in leaf sheath cum stem and 0.6 mg g^{-1} in ear head under drought stress. The proportion of increased total phenol content by the addition of FA + SSB + FYM with 100% STBK under both drought and flood stress was observed; however, the magnitude of increase was superior in drought stress compared to flood stress. The total phenol accumulation in the flowering stage showed positive

Table 3 Effect of FA with SSB + FYM with soil test-based K on total and OD phenol content ($mg\ g^{-1}$) at flowering stage of rice under induced drought stress

Treatments	Drought (flowering stage)					
	Total phenol			OD phenol		
	Leaf	Leaf sheath cum stem	Ear head	Leaf	Leaf sheath cum stem	Ear head
Control	1.80	0.40	0.60	0.84	0.28	0.70
FA @ 25 t ha^{-1} + SSB + FYM +0% STBK	1.86	0.67	0.81	1.61	0.60	1.16
FA @ 25 t ha^{-1} + SSB + FYM +25% STBK	1.99	0.75	0.90	1.69	0.68	1.27
FA @ 25 t ha^{-1} + SSB + FYM +50% STBK	2.08	0.81	0.95	1.76	0.74	1.33
FA @ 25 t ha^{-1} + SSB + FYM +75% STBK	2.19	0.87	1.0	1.80	0.79	1.38
FA @ 25 t ha^{-1} + SSB + FYM +100% STBK	2.25	0.91	1.04	1.82	0.82	1.41
SED	0.03	0.02	0.01	0.03	0.006	0.02
CD ($P = 0.05$)	0.08	0.05	NS	0.06	0.01	0.04

and significant correlation with Si and K content. The correlation revealed Si content in leaf blade ($r = 0.69$), leaf sheath cum stem ($r = 0.40$) and in ear head ($r = 0.72$), K content in leaf blade ($r = 0.96$), leaf sheath cum stem ($r = 0.98$) and in ear head ($r = 0.97$). The increase in total phenol content by the addition of FA with SSB and FYM with graded level of 100% STBK might be due to the application of K which affects the activity of several important enzymes such as ATPase and RuBisCO as a result, and it has pronounced effect on protein, starch synthesis, ion and assimilate transport and stomatal movements, hence, total phenol shows positive correlation with antioxidant activity. As the antioxidant activity is increased to prevent photooxidative damage, phenol content also increased. Increase in the phenol concentration increases the peroxidase activity which plays an important role in oxidating enzymes (Kant et al. 1992). Increased activity of these oxidative enzymes indicates a state of high catabolism induced during pathogenesis. This is in good agreement with the findings of Sharafzadeh (2011).

2. ***Ortho*-dihydroxy Phenol** (Tables 3 and 4)

The different treatments with FA + SSB + FYM with 100% STBK showed significant variations in the *Ortho*-dihydroxy (OD) phenol content. The highest OD phenol content of 1.82 mg g^{-1} in leaf blade, 0.82 mg g^{-1} in leaf sheath cum stem and 1.41 mg g^{-1} in ear head was recorded by the addition of FA with SSB and FYM

Table 4 Effect of FA with SSB + FYM with soil test-based K on total and OD phenol content ($mg\ g^{-1}$) at flowering stage of rice under induced flood stress

Treatments	Flood (Flowering stage)					
	Total phenol			OD phenol		
	Leaf	Leaf sheath cum stem	Ear head	Leaf	Leaf sheath cum stem	Ear head
Control	1.66	0.64	0.65	0.67	0.194	0.202
FA @ 25 t ha^{-1} + SSB + FYM +0% STBK	1.92	0.86	0.86	0.82	0.21	0.24
FA @ 25 t ha^{-1} + SSB + FYM +25% STBK	1.95	0.865	0.86	0.242	0.68	0.272
FA @ 25 t ha^{-1} + SSB + FYM +50% STBK	1.98	0.869	0.87	0.242	0.74	0.272
FA @ 25 t ha^{-1} + SSB + FYM +75% STBK	1.99	0.871	0.87	0.243	0.79	0.273
FA @ 25 t ha^{-1} + SSB + FYM + 100% STBK	1.99	0.874	0.87	0.252	0.82	0.273
SED	0.007	0.01	0.01	0.002	0.0003	0.0003
CD ($P = 0.05$)	0.016	0.04	0.03	0.005	0.0006	0.0007

integrated with 100% STBK over control which recorded only 0.84 mg g^{-1} in leaf blade, 0.28 mg g^{-1} in leaf sheath cum stem, 0.70 mg g^{-1} in ear head under drought stress. The proportion of increased OD phenol content by the addition of FA + SSB + FYM with 100% STBK under both drought and flood stress was observed; however, the magnitude of increase was superior in drought stress compared to flood stress. The OD phenol accumulation was more in the flowering stage, and it showed positive and significant correlation with Si and K content at the flowering stage with the correlation coefficient of 0.84, 0.47, 0.84, for Si content in leaf blade, leaf sheath cum stem and ear head, respectively. Similarly, significant and positive correlation coefficient values of 0.8, 0.85, and 0.84 were observed for K content with OD phenol in leaf blade, leaf sheath cum stem, and ear head, respectively. The increase in OD phenol content was reported by the addition of FA with SSB and FYM with graded level of 100% STBK might be due to the activity of several important enzymes such as ATPase and RuBisCO as a result, and it has pronounced effect on protein, starch synthesis, ion and assimilate transport and stomatal movements; hence, ortho-dihydroxy phenol shows a positive correlation with antioxidant activity. As the antioxidant activity is increased to prevent photooxidative damage, phenol content also increased. Increase in the phenol concentration increases the peroxidase activity which plays an important role in oxidating enzymes (Kant et al. 1992). This is in good agreement with the findings of Sharafzadeh (2011).

3. **Reducing sugar** (Fig. 5)

The treatmental effects showed significant variations in the reducing sugar content in different parts of the rice plant at flowering stage. The lowest mean reducing sugar content of 1.07 mg g^{-1} in leaf blade, 1.01 mg g^{-1} in leaf sheath cum stem and 1.2 mg g^{-1} in ear head was recorded by the addition of FA + SSB + FYM; whereas, the control recorded 1.17, 1.06, 1.32 mg g^{-1} in leaf blade, in leaf sheath cum stem and ear head, respectively. The graded levels of STBK showed a significant difference in the reducing sugar content and recorded the lowest mean reducing sugar content of 1.09 mg g^{-1} in leaf blade, 0.97 mg g^{-1} in leaf sheath cum stem and 1.27 mg g^{-1} in ear head by the addition of 100% K; whereas, the control recorded 1.13, 1.08, 1.28 mg g^{-1} in leaf blade, leaf sheath cum stem and ear head, respectively. The interaction effect of treatments showed significant difference in the reducing sugar in leaf blade and ear head, and there is no significant difference which was observed in leaf sheath cum stem. The reducing sugar showed negative and significant correlation with Si content, at flowering stage. The negative correlation of reducing sugars with Si and K was obsered in different part of rice plant.

4. **Non-reducing sugar** (Fig. 6)

The treatmental effects showed significant difference in non-reducing sugar content at flowering stage and recorded the lowest mean non-reducing sugar content 1.34 mg g^{-1} in leaf blade, 1.32 mg g^{-1} in leaf sheath cum stem and 1.34 mg g^{-1} in ear head by the application of FA with SSB and FYM; whereas, the control recorded 1.41, 1.38, 1.39 mg g^{-1} in leaf blade, leaf sheath cum stem and ear head, respectively. The treatment effects showed significant difference due to the addition of graded levels of STBK and recorded the lowest mean non-reducing sugar content of 1.37 mg g^{-1} in leaf blade, 1.34 mg g^{-1} in leaf sheath cum stem and 1.36 mg g^{-1} in

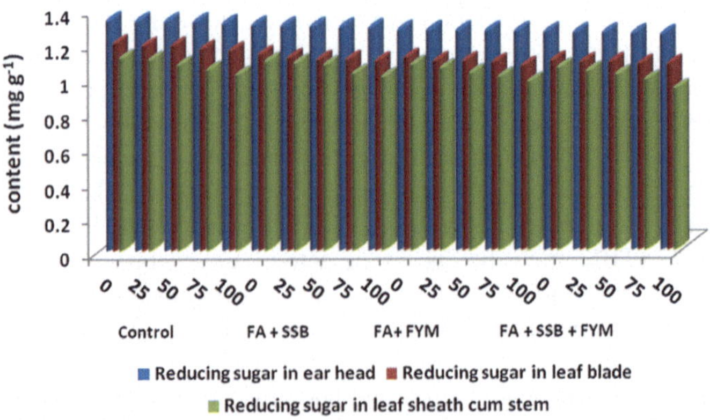

Fig. 5 Effect of fly ash with SSB + FYM with graded levels of STBK on reducing sugar in leaf blade, leaf sheath cum stem and ear head at flowering stage in rice under drought stress

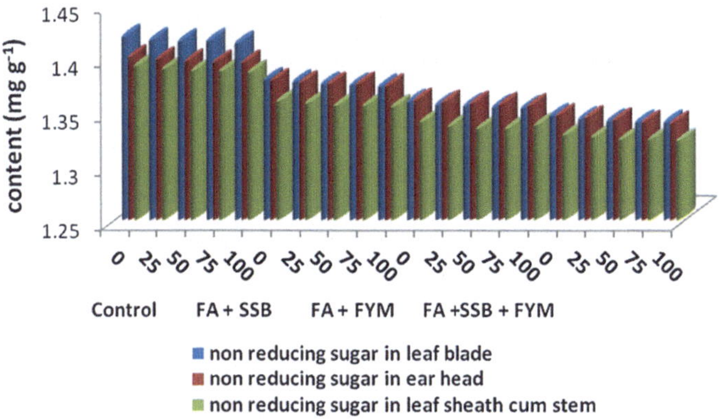

Fig. 6 Effect of fly ash with SSB + FYM with graded levels of STBK on non-reducing sugar in leaf blade, leaf sheath cum stem and ear head at flowering stage in rice under drought stress

ear head by the addition of 100% potassium. The interaction effect of treatments did not show any significant difference in the non-reducing sugar content at the flowering stage. The non-reducing sugar content showed negative and significant correlation with Si and K content, at the flowering stage with the correlation coefficient of −0.99, −0.40, −0.99 for leaf blade, leaf sheath cum stem and ear head, respectively. Similarly, the correlation coefficient of −0.67, −0.69 and −0.68 was observed for K content in leaf blade, leaf sheath cum stem and ear head, respectively. The decreased content of sugars by different treatments might be due to the release of Si and K from FA, and its uptake by rice plant stimulates the photosynthetic activity and markedly suppresses the non-reducing sugars, starch phospharylase and phosphatase enzyme in rice plants (Horsfall and Diamond 1957). The application of potassium leads to a considerable reduction in the non-reducing sugar in the plants. The reduction in the soluble sugars due to potassium application might be due to higher polymerization of sugars and utilization of the non-reducing sugar for the synthesis of phenolic compounds through the shikimic acid pathway. Sugars tend to increase the susceptibility in the plant by providing an extra source of energy for invaders (Mahadevan 1973). The results obtained due to the effect of FA under flooded condition were not well pronounced as that of drought condition. The results were coping up with the findings of Mallika et al. (2000).

5. **Proline** (Table 5)

The highest proline content of 45.01 $\mu m\ g^{-1}$ in leaf blade, 38.09 $\mu m\ g^{-1}$ in leaf sheath cum stem and 42.68 $\mu m\ g^{-1}$ in ear head was recorded due to the addition of FA + SSB + FYM with 100% STBK over rest of the treatments. The control recorded the lowest proline content of 19.9 $\mu m\ g^{-1}$ in leaf blade, 13.41 $\mu m\ g^{-1}$ in leaf sheath cum stem and 19.05 $\mu m\ g^{-1}$ in ear head under drought stress condition. The proline accumulation was more during the flowering stage and it was showed positive correlation with Si and K content with correlation coefficient viz. Si content in leaf blade ($r = 0.64$), Si content in leaf sheath cum stem ($r = 0.40$)

Table 5 Effect of FA with SSB + FYM with soil test-based K on proline content ($\mu m\ g^{-1}$) at flowering stage of rice under induced stress

Treatments	Drought			Flood		
	Leaf	Leaf sheath cum stem	Ear head	Leaf	Leaf sheath cum stem	Ear head
Control	19.90	13.41	19.05	12.53	9.47	19.05
FA @ 25 t ha^{-1} + SSB + FYM + 0% STBK	31.16	24.37	26.36	18.60	14.78	15.88
FA @ 25 t ha^{-1} + SSB + FYM +25% STBK	33.76	27.60	29.39	18.72	14.1	15.93
FA @ 25 t ha^{-1} + SSB + FYM +50% STBK	38.09	28.70	32.36	18.78	14.5	15.97
FA @ 25 t ha^{-1} + SSB + FYM +75% STBK	41.12	32.89	37.32	18.80	14.98	16.0
FA @ 25 t ha^{-1} + SSB + FYM + 100% STBK	45.01	38.09	42.68	18.82	15.01	16.02
SED	1.33	1.07	0.93	0.008	0.02	0.02
CD	NS	2.27	1.96	NS	NS	NS

and in ear head (r = 0.65) and K content in leaf blade (r = 0.63), leaf sheath cum stem (r = 0.39) and in ear head (r = 0.62). On comparing before and after drought stress, the more accumulation of proline was observed after induced drought stress. Proline often referred as "compatible solutes" is the most water-soluble amino acids and exists much of the time in a zwitterionic state having both weak negative and positive charges at the carboxylic acid and nitrogen groups, respectively. Proline concentrations increase under lower water availability and higher Si availability, which indicate that Si may be associated with plant osmotic adjustment. Proportionately, a similar trend of results was obtained under induced flood stress condition; however, the effect was less pronounced in response to the accumulation of proline. The results were corroborated with the findings of Ma et al. (2001), Ma and Takahashi (2002), Hattori et al. (2005) and Gao et al. (2006).

4 Conclusions

Application of FA @ 25 t ha^{-1} + SSB @ 2 kg ha^{-1} + FYM @12.5 t ha^{-1} with 100% soil test-based potassium improved the yield of rice. This treatment enhanced the uptake of silicon in both grain and straw at maturity. It also helps in mitigating the biotic stresses, viz. the incidence of stem borer and ear head bug in rice by the influence of Si under induced abiotic stress, viz. drought and flood. Other biochemical constituents, viz. phenolics, sugars and proline were also favorably correlated with the reduced incidence of pests under drought and flood stress situation.

Acknowledgements The authors are grateful to the Authorities of the FA Unit, Department of Science and Technology, Government of India, New Delhi, for providing financial assistance to carry out the above investigations.

References

Arivazhagan K., Ravichandran, M., Dube, S. K., Mathur, V. K., Khandakar, R. K., Tripati, V. K. M., et al. (2011). Effect of coal fly ash on agricultural crops: Showcase project on use of fly ash in agriculture in and around thermal power station areas of National Thermal Power Corporation Ltd. In *World of Coal Ash conference*, Denver, USA.

Balasubramaniam, P. (2003). *Studies on the utilization of rice straw as a source of silicon and potassium for low land rice in Udic haplustalf.* Ph.D thesis, TNAU, Coimbatore.

Chandramani, P., Rajendran, R., Sivasubramanian, P., & Muthiah, C. (2009). Management of hoppers in rice through host nutrition—A novel approach. *Journal of Biopesticides, 2*(1), 99–106.

Chandramani, P. (2010). Organic sources induced silica on leaf folder, stem borer and gall midge population and rice yield. *Journal of Biopesticides, 3*(2), 423–427.

Djamin, A., & Pathak, M. D. (1967). Role of silica in resistance to Asiatic rice borer, *Chilo suppressalis* (Walker) in rice varieties. *Journal of Economic Entomology, 60,* 347–351.

El-Hadi, A. H. A., Ismail, K. M., & Akahawy, M. A. (1997). Effect of Potassium on the drought resistance of crops in Egyptian conditions. In A. E. Johnston (Ed.), *Food security in the WANA region, the essential need for balanced fertilization* (pp. 328–336). Basel Switzerland: International Potash Institute.

Gao, X., Zou, G. H., Wang, L., & Zhang, F. (2006). Silicon decrease transpiration rate and conductance from stomata of maize plants. *Plant Nutrition, 29,* 1637–1647.

Horsfall, J. G., & Diamond, A. E. (1957). Interactions of tissue sugar, growth substances and disease susceptibility. *Plant Pathology, 64,* 415–421.

Hattori, T. S., Inanaga, H., Araki, P., Morita, S., Luxova, M., & Lux, A. (2005). Application of silicon enhanced drought tolerance in *Sorghum bicolor*. *Physiologia Plantarum, 123,* 459–466.

Hong, C. O., Lee, C. H., Lee, Y. B., & Kim, P. J. (2006). Evaluating possibility of heavy metal accumulation by fly ash application in rice paddy soils. *Korean Journal of Environmental Agriculture, 25*(4), 331–338.

Jackson, M. L. (1973). *Soil chemical analysis*. New Delhi, India: Prentice Hall of India (p) Ltd.

Jones, L. H. P., & Handreck, K. A. (1967). Silica in soil, plants and animals. *Advances in Agronomy, 19,* 107–149.

Kant, U., Qureshi, M. A., & Jain, P. (1992). In vitro studies of nematode induced gall of economically important plants. In T. Singh & P. C. Trevedi (Eds.), *Vistas of seed biology* (pp. 91–97).

Lee, H., Sung, C. H., Lee, Y. B., & Kim, Leeand P. J. (2006). Fly ash effect on improving soil properties and rice productivity in Korean paddy soils. *Bioresource Technology, 97,* 1490–1497.

Lindsay, W. L. (1979). *Chemical equilibria in soil* (pp. 499). New York: Wiley.

Ma, J. F., & Takahashi, E. (2002). A text book on *Soil, fertilizer, and plant silicon research in Japan* (pp. 267). Amsterdam, The Netherlands: Elsevier.

Ma, J. F., Goto, S., Tamai, K., & Ichii, M. (2001). Role of root hairs and lateral roots in silicon uptake by rice. *Plant Physiology, 127,* 1773–1780.

Mahadevan, A., & Sridhar, R. (1986). *Methods in physiological plant pathology* (3rd ed.). Indra Nagar, Chennai: Sivakami Publications.

Mahadevan, A. (1973). Theoretical concepts of disease resistance. *Acta Phytopathology, 8,* 391–423.

Mallika, S., Ramabadran, R., & Balasubramaniam, P. (2000). Effect of lignite fly ash on yield and its attributes on healthy and blast inoculated rice. *ORYZA, 37*(4), 342–343.

Maxwell, F. G., Jenkins, J. N., & Parroot, W. L. (1972). Resistance of plants to insects. *Advanced Agronomy, 24,* 187–265.

Mengel, K., & Kirkby, E. A. (2001). *Principles of plant nutrition* (5th ed.). Dordrecht: Kluwer Academic Publishers.

Meyer, J. H., & Keeping, M. G. (2005). Impact of Silicon in alleviating biotic stress in sugarcane in South Africa. In *Proceedings of the International Society of Sugar Cane Technologists.*

Narayanasamy, P. (1994). *Studies on use of lignite flyash as an insecticide and an adjuvant in insecticide formulations.* Final report of the scheme titled. Chennai: Tamil Nadu State Council for Sciences and Technology.

Panda, N., Pradhan, B., Samalo, A. P., & Rao, P. S. P. (1977). Note on the relationship of some biochemical factors with resistance in rice varieties to yellow borer. *Indian Journal of Agricultural Science, 45,* 499–501.

Salim, & Saxena, (1992). Iron, silica, and aluminum stresses and varietal resistance in rice: Effects on white backed plant hopper. *Crop Science, 32*(1), 212–219.

Sarwar, G., Schmeisky, H., Hussain, N., Muhammad, S., ibrahim, M., & Ehsan, S. (2008). *Improvement of soil physical and chemical properties with compost application in rice-wheat cropping system. Pakistan Journal of Botany, 40*(1), 275–282. Department of Soil and Environmental Sciences. University of Sargodha, Sargodha, Pakistan. Witzenhausen,Germany. Soil Salinity Research Institute, Pindi Bhattian, Pakistan.

Sharafzadeh, S. H. (2011). Effect of nitrogen, phosphorous and potassium on growth, essential oil and total phenolic content of garden thyme (*Thymus vulgaris* L.). *Advances in Environmental Biology, 5,* 699–703.

Tayabi, K., & Azizi, P. (1984). Influence of silica on rice yield and stem borer (Chilosupremani) in result) Iranin 1979–1980. *Pesticides, 18,* 20–22.

The manufacturer's authorised representative in the EU is Springer Nature Customer Service Centre GmbH, Europaplatz 3, 69115 Heidelberg, Germany. If you have any concerns regarding our products, please contact ProductSafety@springernature.com

Printed and bound by CPI Group (UK) Ltd, Croydon, CR0 4YY
25/09/2026
02222900-0017